Drew Abbott

WILLFORD PRESS

www.willfordpress.com

Published by Willford Press,
118-35 Queens Blvd., Suite 400,
Forest Hills, NY 11375, USA

ISBN: 978-1-64728-497-8

Cataloging-in-Publication Data

Construction materials, equipment and techniques / Drew Abbott.
p. cm.
Includes bibliographical references and index.
ISBN 978-1-64728-497-8
1. Building materials. 2. Civil engineering. I. Abbott, Drew.
TA403.6 .M38 2023
620.11--dc23

For information on all Willford Press publications
visit our website at www.willfordpress.com

WILLFORD PRESS

Contents

Preface... IX

Chapter 1   **Concrete Technology**.. 1

     1.1 Cements.. 1

     1.2 Concrete Chemicals and Applications.................................. 5

     1.3 Grade of Concrete... 10

     1.4 Manufacturing of Concrete... 11

     1.5 Compaction of Concrete.. 13

     1.6 Curing and Finishing 14

     1.7 Testing of Fresh and Hardened Concrete 17

     1.8 Quality of Concrete.. 28

     1.9 Extreme Weather Concreting... 29

     1.10 Ready Mix Concrete 30

     1.11 Non-Destructive Testing.. 32

Chapter 2   **Construction Practices**.. 37

     2.1 Specifications, Details and Sequence of Activities............ 37

     2.2 Construction Co-Ordination 40

     2.3 Site Clearance 40

     2.4 Marking... 41

     2.5 Earthwork 43

     2.6 Masonry 49

     2.7 Concrete Hollow Block Masonry.. 54

     2.8 Flooring 57

     2.9 Damp Proof Courses 68

     2.10 Construction Joints... 73

     2.11 Movement and Expansion Joint 76

     2.12 Precast Pavements.. 77

     2.13 Building Foundations... 78

     2.14 Basements 82

     2.15 Temporary Shed.. 84

     2.16 Centering and Shuttering 85

     2.17 Slip Forms 86

     2.18 Scaffoldings 87

2.19 De-Shuttering Forms 92

2.20 Fabrication and Erection of Steel Trusses 93

2.21 Frames 96

2.22 Braced Domes 100

2.23 Laying Brick 100

2.24 Weather and Water Proof 102

2.25 Roof Finishes 104

2.26 Acoustic and Fire Protection 108

Chapter 3   **Sub-Structure Construction** **116**

3.1 Techniques of Box Jacking 116

3.2 Pipe Jacking 117

3.3 Under Water Construction of Diaphragm Walls and Basement 118

3.4 Tunneling Techniques 120

3.5 Piling Techniques 125

3.6 Well and Caiss 126

3.7 Sinking Cofferdam 131

3.8 Cable Anchoring and Grouting 132

3.9 Driving Diaphragm Walls and Sheet Piles 135

3.10 Shoring for Deep Cutting 141

3.11 Well Points 144

3.12 De-Watering and Stand by Plant Equipment for Underground
       Open Excavation 148

Chapter 4   **Super-Structure Construction** **151**

4.1 Launching Girders, Bridge Decks and Off Shore Platforms 151

4.2 Special Forms for Shells 155

4.3 Techniques for Heavy Decks 159

4.4 In-Situ Prestressing in High Rise Structures 162

4.5 Material Handling 165

4.6 Erecting Light Weight Components on Tall Structures 167

4.7 Support Structure for Heavy Equipment and Conveyors 171

4.8 Erection of Articulated Structures, Braced Domes and Space Decks 173

Chapter 5   **Construction Equipment** **175**

5.1 Selection of Equipment for Earth Work 175

5.2 Earth Moving Operations 177

5.3 Types of Earthwork Equipment 178

5.4 Equipment for Foundation and Pile Driving.. 183

5.5 Equipment for Compaction, Batching, Mixing and Concreting............................. 187

5.6 Equipment for Material Handling and Erection of Structures................................ 192

5.7 Equipment for Dredging, Trenching and Tunneling.. 198

Permissions

Index

Preface

The purpose of the book is to provide a glimpse into the dynamics and to present opinions and studies of some of the scientists engaged in the development of new ideas in the field from very different standpoints. This book will prove useful to students and researchers owing to its high content quality.

Construction is a broad term that refers to the activities related to the building of large or small structures, objects and systems such as school buildings, bridges, dams and tunnels. It comprises functions such as designing, planning and layout, construction financing, and implementing the design. Construction materials refer to all the materials that are utilized in the construction of a structure. Many naturally occurring materials, including clay, rocks, sand, wood, and even twigs and leaves have been used to build structures. Several manufactured items, such as cement, bricks, glass, metal and plastics are also used in addition to the naturally existing materials. Bulldozers, cranes, excavators, dump trucks, and loaders are some of the most common types of construction equipment. There are various types of construction techniques including light gauge steel construction, hybrid concrete construction, 3D volumetric construction, flat slabs, thin-joint masonry, and joisted or load bearing masonry construction. This book is compiled in such a manner, that it will provide in-depth knowledge about the materials, equipment and techniques related to construction. It will prove to be immensely beneficial to students and researchers in this field.

At the end, I would like to appreciate all the efforts made by the authors in completing their chapters professionally. I express my deepest gratitude to all of them for contributing to this book by sharing their valuable works. A special thanks to my family and friends for their constant support in this journey.

Drew Abbott

Concrete Technology

1.1   Cements

Hydraulic cement, commonly known as cement is one of the most extensively used basic material in almost all civil engineering constructions. Also referred to as Portland cement or Ordinary Portland Cement (OPC).

It is a finely ground material with the addition of requisite quantity of water, it is capable of hardening both under water and in air by the chemical interaction of its constituents with water, it is capable of building together with appropriate materials. Consists of two main constituents, viz., argillaceous materials-main ingredient is clay and calcareous materials-main ingredient is calcium carbonate.

Constituents of Ordinary Portland Cements

Various constituents (ingredients) of an Ordinary Portland Cement are:

- Lime (62 %).

- Silica (22 %).

- Alumina (5 %).

- Calcium Sulphate (4 %).

- Iron Oxide (3 %).

- Magnesium Oxide (2 %).

- Sulphur Trioxide (1 %).

- Alkalies (1 %).

Chemical Composition of Cement

Raw materials used for the formation of cement consist mainly of lime, silica, alumina and iron oxide. In the kiln, at high temperature the constituents of cement interact with one another.

Oxides of these materials in proper quantities are responsible for influencing the various properties of cement including the rate of cooling and fineness of grinding. Raw materials at high temperature combine with each other and form complex compounds.

Four compounds usually regarded as major compounds are:

- Tricalcium silicate $(3 \ CaOSiO_2 \ or \ C_3S)$.

- Dicalcium silicate $(2 \ CaOSiO_2 \ or \ C_2S)$.

- Tricalcium aluminate $(3 \ CaOAl_2O_3 \ or \ C_3A)$.

- Tetra-calcium aluminoferrite $(4 \ CaOAl_2O_3Fe_2O_3 \ or \ C_4AF)$.

In addition to the above four compounds, minor compounds $(K_2O \ and \ N_2O)$ are also formed which are not significant. Tricalcium silicate and dicalcium silicate are the most important compounds which contribute to the strength. Presence of free lime in the clinker will cause unsoundness in the cement. High total alumina ferric oxide produces high early strength.

Properties of Cement

- It is one of the best binding materials used in civil engineering constructions.

- It has high plastic properties.

- It offers high strength to masonry.

- It hardens in short time.

- It has high resistance to water and other atmospheric effects.

Quality Requirements of Cement

- It should be homogeneous and be uniform in color.

- It should be free from lumps and should sink in water if a small quantity is placed on the surface of water.

- The ratio of percentage of alumina to that of iron oxide should not be less than 0.66.

- The weight of magnesia should not exceed 5 %.

- The total sulphur content of cement should not be greater than 2.75 %.

- The initial setting time should not be less than 30 minutes and the final setting time should be around 10 hours.

Types of Cement and Uses

Different types of cements are manufactured by changing the chemical composition and using different raw materials and additives to suit the specific need:

1. Ordinary Portland Cement

Common type of cement which is used for construction of many structures in the form of mortar and concrete, viz., multistory buildings, dams, bridges, storage reservoirs, residential buildings, roads, runways, etc. Used for making joints for pipes, manufacture of precast pipes, piles, hollow block bricks, etc.

2. High Alumina Cement

It is the cement obtained by grinding high alumina clinker. It has long initial setting time, high ultimate strength, high resistance to the action of acids and high temperature. It is used for furnace insulation, refractory concrete and for special structures which require corrosion resistance.

3. Portland-Pozzolana Cement

It is an intimately inter ground mixture of Portland clinker and pozzolana with the possible addition of gypsum or an intimate and uniform blend of Portland cement and fine pozzolana. It takes more time for initial setting which helps in works which involve delayed construction.

4. Masonry Cement

It is a product obtained by inters grinding a mixture of Portland cement clinker with inert materials (non-pozzolanic) and gypsum and air entraining plasticiser. This type is characterised by certain physical properties, such as slow-hardening, high workability and high water retentivity which makes it especially suitable for masonry work.

5. Hydrophobic Cement

It is obtained by grinding Ordinary Portland Cement clinker with an additive and will impart a water repelling property. It shall be destroyed only by wet attrition, such as in concrete mix. The hydrophobic quality of cement would facilitate its storage for longer periods in extremely wet climate conditions.

6. Oil-Well Cement

Hydraulic cement which contains retarders in addition to coarse grinding and/or reduced Tri calcium aluminate content of clinker. Suitable for use in high pressure and temperature in sealing water and gas pockets and setting, causing during the drilling and repair of oil-wells.

7. Quick-Setting Cement

Produced by adding a certain quantity of aluminium sulphate and reducing the quantity of gypsum and a fine grinding is made. Used for under-water concreting.

8. Low-heat Cement

Contains low quantity of tricalcium aluminate and high quantity of dicalcium silicate. Used for mass concreting of dams.

9. Expanding Cement

Obtained by adding an expanding medium like sulpho-aluminate and a stabilising agent for ordinary cement. Unlike the conventional cement of shrinking, it expands during curing. Used for repairing concrete surfaces.

10. Rapid Hardening Cement

Produced by burning the raw materials at high temperature and by increasing lime content. It has the quality of attaining high strength in a short period. Used in the works wherein speed of construction is needed.

11 Acid-resistant Cement

Materials like quartz, sodium silicate and sodium fluosilicate are added to the cement to attain the acid-resistant quality which is used in chemical industry.

12. Sulphate-resistant Cement

It has higher silicate content which is effective in fighting back the attacks of sulphates. Tricalcium content is restricted to 5 % only. It has high resistance to sulphate. Used for construction of sewage treatment works, marine structures and foundation in soil with high sulphate.

13. White Cement

It does not contain colouring ingredients such as iron oxide, manganese oxide or chromium oxide. White cement is burnt with oil. It is used for floor finish, plastering, pointing of masonry, manufacture of precast stones, tiles and colour cement and runway markings.

14. Coloured Cement

The required colour of the cement, can be obtained by initial mixing, colouring materials with the cement. Used for external finishing of walls and floors, manufacturing of tiles and precast stones and also used for paths, swimming pools and tennis courts.

1.1.1 Grade of Cements

The OPC has been classified into three grades, namely 33 grade, 43 grade and 53 grade based upon the strength of the cement. If the strength is not less than 33 N/mm², it is called 33 grade cement. If the strength is not less than 43 N/mm², it is called 43 grade cement. If the strength is not less than 53 N/mm², it is called 53 grade cement.

Type of cement	Fineness	Soundness mm	Setting time		28 Days compressive strength Mpa
			Initial	Final	
			min	min	
33 Grade OPC	225	10	30	600	33
43 Grade OPC	225	10	30	600	43
53 Grade OPC	225	10	30	600	53

33 grade OPC:

It is used for normal grade of concrete up to M20, plastering, flooring, grouting of cable ducts in PSC works etc. The fineness should be in the range of 225 and 280.

43 grade OPC:

- It is the most widely used general purpose cement. For concrete grades up to M30, precast elements.

- It is used for marine structures but C3A should be in the range of 5 – 8%.

53 grade OPC:

- For concrete grade higher than M30, PSC works, bridge, roads, multi-storied buildings etc.

- For use in cold weather concreting.

- It is also used in marine structures but C3A should be between 5 – 8%.

1.2 Concrete Chemicals and Applications

Admixtures and construction chemicals are the chemicals added along with the ingredients of concrete or afterwards to get the required mix to fit in for the desired strength and durability.

Admixtures

Concrete is used for varied purposes to make suitable for different occasions and in environments. Ordinary concrete does not fit in for varied purposes. Additives are the materials which are added at the time of grinding cement clinker in the cement factories. Effect of admixtures depends on the brand of cement, grading of aggregate, mix proportion and richness of the mix. With caution of the admixtures should be selected in correctly predicting the behaviour of concrete.

Important Admixtures

There are several admixtures available, some important of them are discussed below:

1. Plasticizers and Super Plasticizers

High degree of workability needs in different situations. Addition of excess water will only help the fluidity and not the workability of concrete. The addition of plasticizers will improve the desirable qualities demanded for plastic concrete.

Plasticizers are based on the following constituents:

- Anionic surfactants such as ligno sulphonates and their modifications.
- Nonionic surfactants, such as poly glycol acid of hydroxylated carboxylic acids and their modifications.
- Others such as carbohydrates, etc.

Among the plasticizers, calcium, sodium and ammonium ligno sulphonates are mostly used. The quantity used are 0.1 to 0.4 % by weight of cement. Super plasticizers constitute relatively a new and improved form of plasticizers.

These are chemically different from conventional plasticizers. Special quality of super plasticizers are the powerful action as dispersing agents and they are high range water reducers. Are chemically different from plasticizers. Super plasticizers permit reduction of water up to 30 % without reduction in workability.

Super plasticizers are used for production of flowing, self-leveling and self-compacting and for production of high strength and high performance concrete.

2. Retarders

Retarders are an admixture which slows the process of hydration. Concrete remains plastic and workable because of retarders. If concreting is done in hot weather, retarders overcome the accelerating effect of high temperature. Retarders are used in consolidating a large number of pours without the formation of cold joints and in grouting oil wells. Calcium sulphate is the commonly used retarder.

Admixtures increase the compressive strength by 10 to 20 %. Retarding plasticizers are available in the market. These are an important type of admixtures often used in the ready mixed concrete industry, for retaining the slump loss, during high temperature, long transportation and to avoid construction cold joints.

3. Accelerators

It is a useful type of admixture which is added to get the early strength. Such situations may occur under the following conditions:

- When early removal of form work is needed.
- When reduction of period of curing is needed.
- When to put the structure early to use.
- When accelerating the setting time in cold weather.
- For emergency repair work.

Commonly used accelerator was calcium chloride. Soluble carbonates, silicates, fluosilicate and some of the organic compounds are used. Fluosilicates and organic compound like trietheuolamine are comparatively expensive. Some of the accelerators available now can make the cement set into stone hard in a matter of five minutes. The availability of such accelerators for underwater concreting has become easy. Waterfront structures which need repair in short time may be done using accelerators. These materials could be used in cold environments up to 60°C.

4. Air-entraining Admixture

Air-entrained concrete is done using an air-entrained cement or addition of air-entraining agent. Air-entraining agents produce a large quantity of air bubbles which act as flexible ball bearings and modify the properties of concrete regarding workability, segregation, bleeding and finishing quality of concrete. Hardened concrete gains resistance to frost action and permeability. Different air entraining agents behave differently depending on the elasticity of the film of the bubble formed and the extent to which the surface tension is reduced.

5. Pozzolanic Admixtures

Pozzolanic or mineral admixtures have been in use since advent of concrete. Application of pozzolanic modifies certain properties of fresh and hardened concretes. Proper addition of pozzolanic admixtures to cement contributes the following:

- It lowers the heat of hydration.
- It increases the water tightness.

- It reduces the alkali-aggregation reaction.

- It resists sulphate attack & improves workability, etc.

Siliceous materials and aluminous materials, do not possess any cementitious materials. On reacting with cement and moisture, they chemically react with calcium hydroxide liberated on hydration and form compounds possessing cementitious properties. This reaction is called pozzolanic reaction. Naturally available pozzolanic materials are clay and shales, diatomaceous earth, volcanic tuffs and pumicites. Artificially available pozzolanic materials are fly ash, blast furnace slag, silica fume, rice husk ash, metalkaoline and surkhi.

6. Damp-proofing Admixtures

Two important properties, concrete should possess with reference to water are:

- To resist when subjected to presence of water.

- To protect the absorption of surface water by capillary action.

A properly designed and constructed concrete should be impermeable. It has been accepted that the addition of some damp-proofing admixture may prove to be of some advantage in reducing the permeability. Damp-proofing admixtures are available in powder or liquid form. They have the properties of pore filling or water repellant materials. The chief material in pore filling admixtures is a silicate of soda, aluminium and zinc sulphates and aluminium and calcium chloride.

They are also more active material which renders the concrete more impervious and also accelerate the setting time. Hence, mineral oil free from fatty or vegetable oil is also used. Production of low permeability concrete depends on the uniform spreading of the admixture.

Construction Chemicals

Other chemicals which are used to enhance the performance are referred to as construction chemicals or building chemicals.

1. Concrete Curing Compounds

In order to prevent the loss of water, from the surface due to evaporation, it has to be retained for which certain measures are taken calling curing. Surface loss of water from concrete depends upon air temperature, relative humidity, fresh concrete temperature and wind velocity. Liquid membrane forming curing compounds are used. Curing compounds is used with the following bases, viz, synthetic resin, wax, acrylic and chlorinated rubber.

Resin and wax based compounds effectively seal the concrete from surface evaporation. After 28 days of curing these compounds peel off. Acrylic based membrane compounds have the additional advantage of better adhesion of subsequent plaster. Membrane need not be removed, but the plastering can be done over it. Because of the acrylic emulsion the bonding for the plaster is better. Chlorinated rubber curing compounds form a thin film on the surface of the concrete, which protects drying at the same time they fill the pores on the surface of the concrete. Surface film will wear out.

2. Polymer Bonding Agents

New concrete is required to be placed over an old concrete surface. In such cases a perfect bond is required. By providing a bond coat between the new and old surfaces of concrete a bond can be achieved. A mixing of a bonding agent with the new concrete helps to provide a better bond. Such mixtures also improve the workability and reduce shrinkage.

Many types of commercial products such as Roof Bond ERB, Nitobond PVA, etc., are available. Polymer modified repair materials are available for repair of concrete work. Such repair works may be a ceiling of concrete roof, hydraulic structures, prefabricated members, pipes, poles, etc.

3. Water-proofing Chemicals

Many of the admixtures discussed above directly or indirectly reduce the permeability of concrete and thereby making the material water-proof.

However, water-proofing of roofs, walls, bathrooms, toilets, kitchens, basements, swimming pools and water tanks, etc. still pose some problem. Different materials are available to make the concrete perfect waterproof.

They are integral waterproofing compounds, acrylic based polymer, mineral based polymer, chemical DPC, waterproofing adhesive for tiles, silicon based water repellent materials, injection grout, joint sealants and protective and decorative coatings.

S. No.	Group	Grade designation	Specified Characteristic Compressive strength of 150mm cube at 28 Days in N/ mm^2
1.	Ordinary	M10	10
2.	Concrete	M15	15
3.		M20	20
4.	Standard	M25	25
5.	Concrete	M30	30
6.		M35	35
7.		M40	40
8.		M45	45

9.		M50	50
10.		M55	55
11.	High	M60	60
12.	Strength	M65	65
13.	Concrete	M70	70
14.		M75	75
15.		M80	80

1.3 Grade of Concrete

Table: Grades of cement concrete.

Grade of concrete	Total quantity of dry aggregates by mass per 50kg of cement to be taken as the sum of the individual masses of fine and coarse aggregates Maximum (kg)	Proportion of fine aggregate to coarse aggregate (By Mass)	Quantity of water per 50kg of cement maximum (Liters)
M 5	800	Generally 1:2	60
M 7.5	625	But subject to	45
M 10	480	An upper limit	34
M 15	350	1:1 1/2 and a lower limit	32
M 20	250	1:2 1/2	30

- In the designation of concrete mix, M refers to the mix and number to the specified compressive strength of 150 mm size cube at 28 days, expressed in N/mm^2.

- For concrete of compressive strength greater than M55 design parameters given in the standard may not be applicable and the values may be obtained from experimental results.

Characteristic strength is defined as the strength of material below which is not more than 5% of the results are expected to fall. Mix proportioning should be selected to ensure that the workability of fresh concrete is suitable for conditions of handling and placing. After compaction, it should surround all reinforcement and completely fill the formwork when concrete is hardened, it should have the required strength, durability and surface finish. Proportion of cement, aggregates and water is called the mix which has to be decided to get the desired strength.

There are two approaches, viz.:

- By designing the concrete mix called the designed mix.

- By adopting nominal concrete mix called the nominal mix.

The concrete obtained by the process of selecting the required ingredients of concrete and finding their relative proportions with the aim of producing an economical concrete of certain strength and durability is called designed mix.

1.4 Manufacturing of Concrete

In the manufacture of concrete, it should be ensured that every batch of concrete has the same proportions. This is mandatory requirement so as to satisfy two aspects, viz., workability and uniform strength.

Steps Involved in Manufacturing of Concrete

In the manufacture of concrete, the following steps are followed:

- Proportioning of concrete.
- Batching of materials.
- Mixing of concrete.
- Conveyance of concrete.
- Placing of concrete.
- Compaction of concrete.
- Curing of concrete.

Proportioning

The process of relative proportions of cement, sand, coarse aggregate and water, so as to obtain a concrete of desired quality is known as the proportioning of concrete.

Batching

After fixing the desired proportion the quantity of required ingredients, viz, cement, coarse-aggregate, fine-aggregate, have to be measured out in batches for mixing. The process of measuring out ingredients is called batching. It may be done by weight or by volume. Volume batching is inferior to weigh batching as the former one is liable to change of volume in sand, in bulking or aggregate constant void feasibility.

Weigh batching: Here all the ingredients of concrete are directly weighed in kilogram. As the weight of cement bag is 50 kg, 20 bags are needed for 1 tonne of cement. This is a slow process.

Volume batching: Here two units of measurement, viz., liquids are measured in litres and solid materials in cubic metres. All ingredients, viz., water, cement, sand and coarse aggregates are measured in litres, the end product concrete is measured in cubic metres. Here cement is taken as the base and other quantities are measured.

Considering 1 litre of cement equal to 1.44 kg, 50 kg bag of cement has a volume of 3.5 litres. For measuring aggregates wooden boxes with inner volume of 3.5 litres has to be used. Size of box of 40 cm × 35 cm × 25 cm satisfies this 3.5 litres volume requirement. Handles are to be provided on the sides for handling.

Mixing

Mixing can be done by hand or by machine. It should be done thoroughly so as to have a uniform distribution of ingredients which can be judged by uniform color and consistency of concrete.

On a clean, hard and water-tight platform, cement and sand are mixed dry using shovels until the mixture shows a uniform color. Aggregates are then added and thoroughly mixed using shovel until the ingredients are uniformly mixed.

Based on the water-cement ratio, the quantity of water required is calculated and added to the dry-mix. Mixing by machine is always preferred. Concrete mixers are used for mixing concrete and are of two types, viz., continuous mixers or batch mixers.

Continuous Mixers: These are used in works where large quantities of concrete is needed such as dams, bridges, etc.

Batch Mixers: It is also called as drum mixers which consist of drums with blades or baffles inside the drum are rotated. Here all required materials in correct quantity are fed into the hopper of the revolving drum. When the mix has attained adequate consistency, the mix is discharged from the drum and conveyed to the concreting yard.

Transporting

Mixed concrete should be conveyed to the concreting yard as early as possible but within setting into initial setting time. Choice of conveyance depends on several factors, viz., nature of work, distance from mixing place to construction site, height to be lifted, type of cement, etc.

During transit from the point of mixing to the point of placement, the following factors have to be born in mind:

- Care should be taken not to allow segregation of aggregates.

- Containers of the drum should be tight such that there is a minimum loss of water.

- Mixed concrete should be brought to the site before the setting in of initial setting time of the cement.

For ordinary simple works, a temporary ladder is formed to convey the concrete using baskets or pans from hand to hand, i.e., by means of manual labour. For larger and important works, various mechanical devices such as vertical hoists, lift-wells for tall structures, wheelbarrows, etc. are used.

Placing

Before any concrete is placed, the entire placing program is planned. No concrete is placed until formwork is inspected and found suitable for placement. Concrete is placed in its final position before the cement reaches its initial set. Concrete is compacted in its final position within 30 minutes of leaving the mixer and once compacted it should not be disturbed.

In all cases, the concrete is deposited in its final position and should not be re-handled or caused to flow in a manner which may cause segregation, loss of materials, displacement of reinforcement, shuttering or embedded inserts or impair its strength. For locations where direct placement is not possible and in narrow forms, suitable drop and Elephant Trunks to confine the movement of concrete is provided.

Special care is taken where concrete is dropped from a height, especially if reinforcement is in the way particularly in columns and thin walls. Concrete should be placed in the shuttering by shovels or other methods and should not be dropped from a height more than one meter or handle in a manner which will cause segregation.

Concrete placed in restricted forms by borrows; buggies, cars, sort chutes or hand shoveling should be subjected to the requirement for vertical delivery of limited height to avoid segregation and should be deposited as nearly as practicable in its final position. Concreting once started should be continuous until the pour is completed. Concrete should be placed in successive horizontal layers of uniform thickness ranging from 150 mm to 900 mm.

1.5   Compaction of Concrete

Compaction is a process of expelling the entrapped air. If we don't expel this air, it will result into honey combing and reduced strength. It has been found from the experimental studies that 1% air in the concrete approximately reduces the strength by 6%.

Methods of Compaction of Concrete:

- Vibration: To compact concrete we apply energy to it so that the mix becomes more fluid. Air trapped in it can then rise to the top and escape. As a result, the concrete becomes consolidated and we are left with a good dense material that will, after proper curing, develop its full strength and durability.

Vibration is the next and quickest method of supplying the energy. Manual techniques such as rodding are only suitable for smaller projects. Various types of vibrator are available for use on site.

- Poker Vibrators: The poker or immersion, vibrator is the most popular of the appliances used for compacting concrete. This is because it works directly in the concrete and can be moved around easily.

- Sizes: Pokers with diameters ranging from 25 to 75mm are readily available and these are suitable for most reinforced concrete work. Larger pokers are available - with diameters up to 150mm - but these are for mass concrete in heavy civil engineering.

- Radius of Action: When a poker vibrator is operating, it will be effective over a circle centered on the poker. The distance from the poker to the edge of the circle is known as the radius of action. However, the actual effectiveness of any poker depends on the workability of the concrete and the characteristics of the vibrator itself. As a general rule, the bigger the poker and the higher its amplitude, the greater will be the radius of action. It is better to judge from our own observations, as work proceeds on site, the effective radius of the poker we are operating on the concrete, we are compacting.

The length of time it takes for a poker vibrator to compact concrete fully depends on:

- The workability of the concrete: The less workable the mix, the longer it must be vibrated.

- The energy put in by the vibrator: Bigger vibrators do the job faster.

- The depth of the concrete: Thick sections take longer.

1.6   Curing and Finishing

Curing of Concrete

The following methods of curing are adopted depending on the type of work:

1. Direct Curing

In this method, water is directly applied to the surface for curing. In this process, the surface is continuously cured by stagnating water, or using moist gunny bags, straws, etc. These methods are used for horizontal surfaces. Vertical surfaces can be cured by covering moist gunny bags or straws.

2. Membrane Curing

In this method, steps are taken to prevent water evaporation from finished concrete surfaces. This is done by covering the surfaces with waterproof papers, polythene papers or by spraying with patented compounds or bituminous layer to form an impervious film on the concrete surface.

3. Steam Curing

This approach is widely used in precast concrete units. Here, the precast units are kept under warm and damp atmosphere of a steam chamber.

4. Direct Electric Curing

Used in many advanced countries in the production of precast components such as channel units, railway sleepers, battery cast large panels, PCC poles, etc. DEC process consist essentially of passing a current through the concrete, between two immersed electrodes. Since, the freshly mixed concrete is an electrical conductor possessing high resistivity, the passing of current through the fresh concrete results in the generation of heat responsible for its accelerated curing.

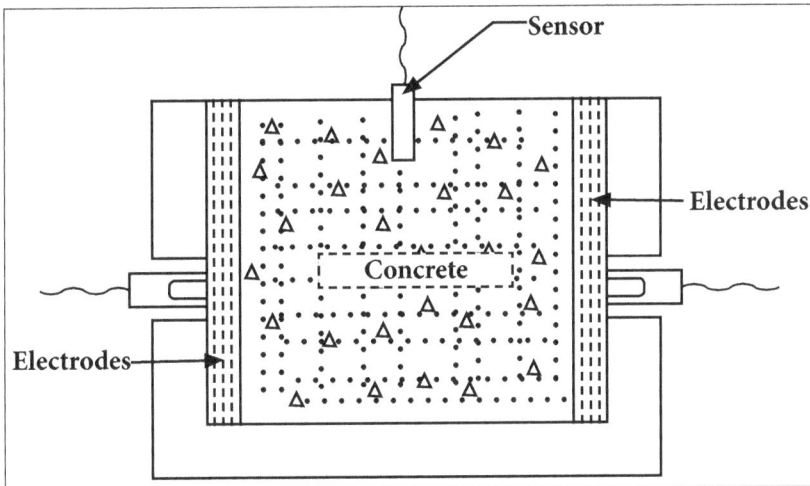

Basic configuration of electrodes.

5. Surface Application by Chemicals

Chemicals such as calcium chloride are spread as a layer on the finished concrete. Chemical absorbs moisture from the atmosphere and prevents evaporation of moisture from the concrete surface.

Accelerated Curing Method

Accelerated curing method is utilized to get early high compressive strength in concrete. This method is also used to find out 28 days compressive strength of concrete in 28 hours.

Accelerated curing is useful in the prefabrication industry wherein high early age strength enables the removal of the formwork within 24 hours thereby decreasing the cycle time resulting in cost saving benefits. The most generally adopted curing techniques are steam curing at atmospheric pressure, warm water curing, boiling water curing and autoclaving.

Procedure:

- Prepare the specimen and store it in moist air of at least 90% relative humidity and at a temperature of 27+2°C for 23 hrs + 15 minutes.

- Lower the specimen, into a curing tank with water at 100 0C and keep it totally immersed for 3 ½ hours + 5 minutes.

- The temperature of water shall not drop more than 30C after the specimens are placed and should return to boiling within 15 minutes.

- After curing for 3 ½ hours + 5 minutes in the curing tank, the specimen shall be removed from the moulds and cooled by immersing in cooling water 27+2°C for a period of at least one hour.

- Read my post compressive strength test of concrete for further steps.

- The corresponding strength at 28 days can be found out from the following correlation:

R_{28} (Strength at 28 days) = 8.09 + 1.64 Ra

Where,

Ra = Accelerated Curing Strength in MPa.

Finishing

Finishing is the last stage in concrete construction. After casting of a concrete, the concrete does not offer a pleasant architectural appearance. In some cases like beams a finishing may not be needed. For a residential building, airport or road pavement and culvert and bridges, the finishing is a must. Many of the prefabricated concrete units are made in such a way to give an attractive architectural effect. Different types of finishes have been in use.

Surface finishes may be grouped as:

- Formwork Finishes.

- Surface Treatment.

- Applied Finishes.

Formwork Finishes

Concrete maintains the shape of formwork, i.e., centering work. Keeping the required shape through formwork, viz., undulated fashionable shapes, V-shaped finishes, plain surfaces or any pleasing surface can be obtained.

Imaginative ideas of architects may be implemented by a careful look of concrete surface. Properly made out formwork can give a very smooth surface using right proportioning of materials better than that made by a best mason. Because of the increasing cost of labour, self-finishing concrete surfaces are preferred.

Surface Treatment

Commonly used method of surface finishing. In order to get a smooth finish, first the proportioning of mix should be appropriate. Finishing of surface should be at the same rate as that of placing of concrete. Adequate care has to be taken to the extent and time of trowelling.

Careful attention should be paid for non-formation of laitance, no excess mortar left and no excess water accumulation on the surface. Rough finishes are required in concrete pavement slab, airfield pavements, in roads, etc. In such cases, the concrete is brought to the plane level surface and then lightly raked, or broomed, or textured or scratched to make the surface rough.

Applied Finish

Exterior application of rendering made on concrete structures is denoted as applied finish. In this case, the concrete surface is finished and kept wet and then a mortar (1:3) is applied. Required pleasant finish is given to the mortar. Rendering applied on wall is pressed with sponge.

By repeating this process, the sand is exposed and the surface gets a finish which is known as Sand Facing. Another type of finish known as a rough cast finish is done. In this type, a wet plastic mix (three parts of cement, one part of lime, six parts of sand and four parts of about 5 mm size of a gravel aggregate) is dashed to the wall surface by means of a scoop or plasterer's trowel. Other finishes under this category are non-slip finish, colored finish, wear resistant floor finish, craziness finish, etc.

1.7 Testing of Fresh and Hardened Concrete

Testing is important in concrete construction. Tests concerned with fresh concrete are to check the workability of concrete. Tests on hardened concrete are to find the strength, creep effects, durability, etc.

Testing of Fresh Concrete

Following tests are commonly employed to measure workability of fresh concrete:

- Slump test.
- Flow test.
- Kelly ball test.
- Compaction factor test.
- Vee Bee consistometer test.

1. Slump Test

Most commonly used method of measuring the consistency of concrete. It can be conducted in the field or in the laboratory. This test is not suitable for very wet or very dry concrete. Apparatus for conducting the slump test consists of a metallic mould in the form of a frustum of a cone with 20 cm bottom diameter, 10 cm top diameter and 30 cm height. Steel tamping rod of 16 mm dia, 0.6 m long with a bullet end is used for tamping. The internal surface of the mould is thoroughly cleaned and placed on a smooth, non-absorbant horizontal surface. Mould is filled with four layers of equal height. Each layer is compacted by giving 25 blows with the tamping rod uniformly. After filling the mould and roding, the excess concrete is stuck off and levelled.

Table: Slump and nature of concrete.

Slump	Nature of concrete mix
No slump	Stiff and extra stiff mix
From 10 to 30 mm	Poorly mobile mix
From 40 to 150 mm	Mobile mix
Over 150	Cast mix

Mould is lifted upwards from the concrete immediately by raising it slowly. This allows the concrete to subside. This subsidence is referred to as slump of concrete. The difference in height of the mould and that of the subsided concrete is measured and reported in mm which is taken as the slump of concrete. Pattern of slump also represents the characteristics of concrete.

If the slump of the concrete is even, it is called a true slump. If one-half of the cone slides down it is called as shear slump. Here, the average value of the slump is considered. Shear slump also indicates that the concrete is not cohesive and reflects segregation.

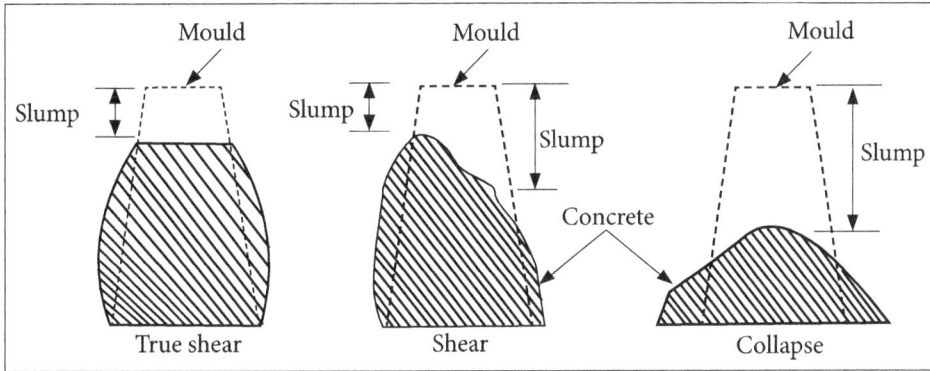

Types of slumps.

Slumps recommended for various works of concrete construction are presented in table:

S. No	Nature of concrete construction	Recommended slump
1.	Concrete to be vibrated	10 to 25 mm
2.	Concrete for road construction	20 to 40 mm
3.	Mass concrete	25 to 50 mm
4.	Concrete for horizontal tops of kerbs, parapets, piers, slabs and walls	40 to 50 mm
5.	Concrete for canal lining	70 to 80 mm
6.	Normal R.C.C. Work	80 to 150 mm
7.	Concrete for arch and side walls of tunnels	90 to 100 mm

Slump test can be conducted both in the laboratory and in work site. Slump test results are used to detect the difference in water content of successive batches of the identical mix.

2. Compacting Factor Test

This is a more refined test than the slump test. Measures the degree of compaction obtained by using certain energy in overcoming the internal friction of the concrete. This property is a measure of workability. The test apparatus consists of two conical hopers with bottom doors and a separate cylinder kept at the bottom. Concrete is filled in the top hopper fully without compaction and released successively through the two hoppers and into the bottom cylinder.

After striking off the level in the cylinder the weight of the concrete (W_1) in the cylinder is determined. Same cylinder is filled with the same batch of concrete and compacted to get the maximum weight (W_2). The ratio of the observed weight (W_1) to the theoretical weight, (W_2), i.e., W_1/W_2 is the compacting factor. The workability, compacting factor and the corresponding slump are given in Table.

Workability	Compacting factor	Corresponding slump
Very low	0.80	0 to 25 mm
Low	0.85	25 to 50 mm
Medium	0.92	50 to 100 mm
High	0.95	100 to 180 mm

This test measures the quality of concrete, which relates very close to the workability. This test clearly depicts the workability of concrete. In the figure, the dimensions are plotted in mm:

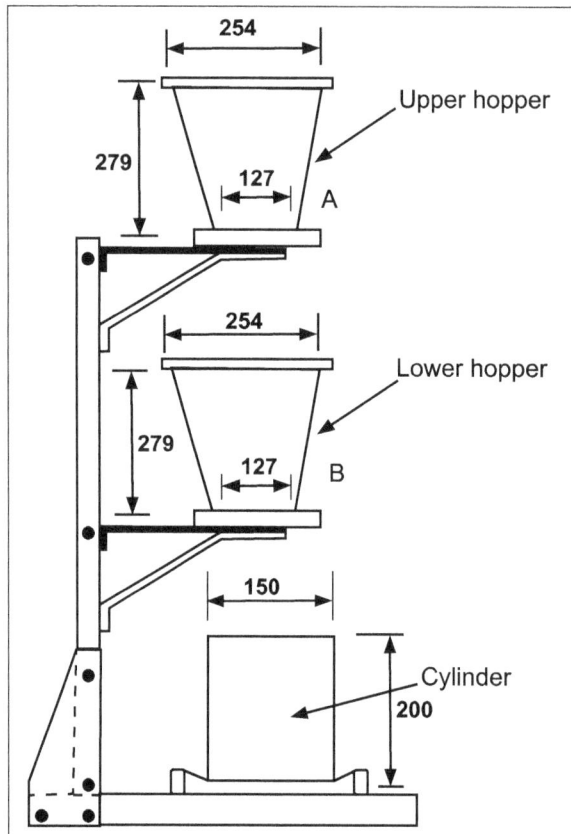

Compacting factor test apparatus.

3. Flow Test

Gives an indication of the quality of concrete with respect to consistency, cohesiveness and non-segregation. Here, a mass of concrete is subjected to jolting and the flow or spread of the concrete is measured. The flow is related to workability.

The test apparatus consists of a flow table of 76 mm dia on which concentric circles are marked. A mould similar to the one used in the slump test with a base diameter is 25 cm and as upper diameter as 17 cm with a height of 12 cm is used.

Mould is kept on the clean table and concrete is filled in two layers with each layer being rodded 25 times with a tamping rod of 1.6 cm diameter and 61 cm long with rounded ends.

Excess concretes on the top of the mould is levelled. The mould is lifted vertically upwards completely. The concrete stands on its own without support.

The table is raised and dropped 12.5 mm with a cam arrangement, 15 times in about 15 seconds. The diameter of the spread-concrete is measured in 6 directions and the average value is taken.

The flow of the concrete is defined as the percentage increase in the average diameter of the spread-concrete to the base diameter of the mould. i.e.,

$$\text{Flow}(\%) = \frac{\text{Average spread in diameter in cm} - 25}{25} \times 100$$

The value varies from 0 to 150 %. Spread pattern of the concrete also reflects the tendency of the segregation. It is a laboratory test. In the figure, the dimensions are plotted in cm:

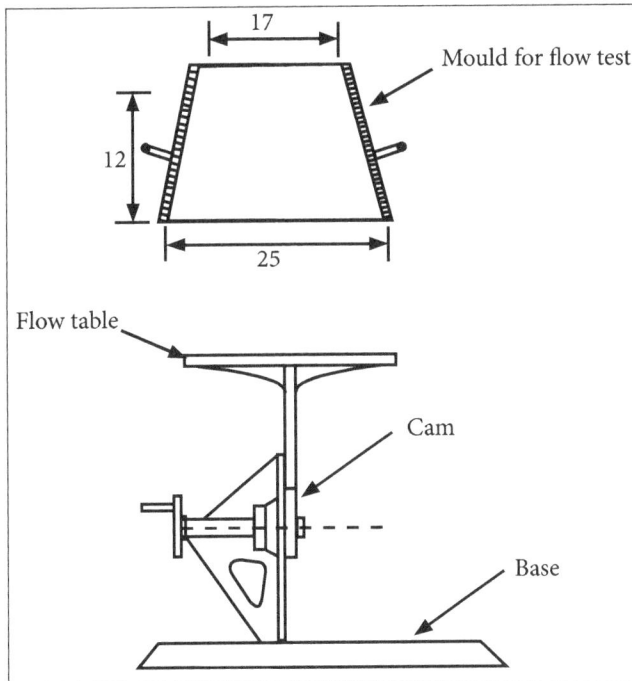

Flow table apparatus.

4. Kelly-Ball Test

Consists of a metal hemisphere of 15 cm diameter, weighing 13.6 kg. Concrete base should be 20 cm depth and the minimum distance from the center of the ball to nearest edge of the concrete is 23 cm. Ball is lowered gradually to the surface of the concrete. Depth of penetration is read immediately on the stem to the nearest 5 mm. This test can be done in a shorter periods of about 15 secs. It gives more consistent results than slump tests. It can be performed in the field and on the concrete placed on the site.

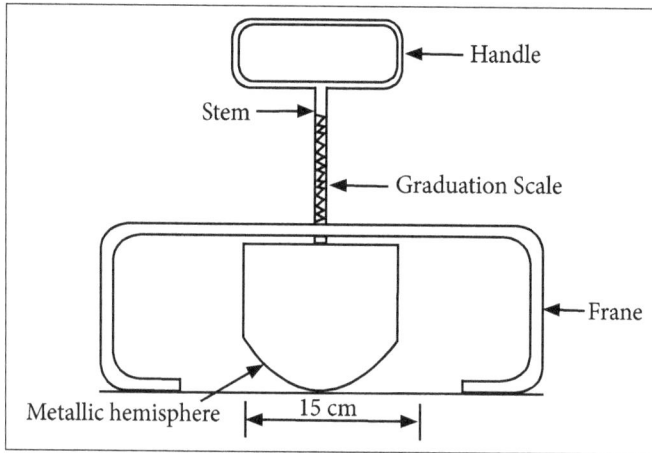

Kelly ball.

5. Vee-Bee Consistometer Test

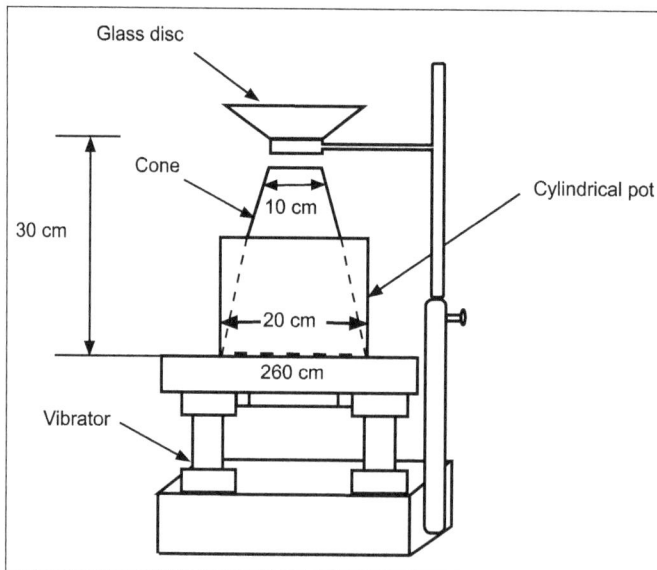

Vee-Bee Consistometer Test.

Consists of a vibrating table, a metal pot, a sheet metal cone and a standard iron rod. A slump cone with concrete is placed inside the sheet metal cylindrical pot of the conistometer. Glass disc is turned and placed on the top of the concrete in the pot. Vibrator switches on and the stopwatch is started simultaneously. Vibrator is kept on till the concrete in the cone assumes a cylindrical shape. The time is noted, time required in seconds for the concrete to change from the shape of the cone to the shape of a cylinder is known as Vee Bee Degree. It is a good laboratory method and more suitable for very dry concrete. Measures the workability indirectly.

Testing of Hardened Concrete

The following tests are conducted for hardened concrete:

- Compressive strength test.
- Flexural strength test.
- Split-tension test.

1. Compressive Strength Test

This is an important test. Most of the properties of concrete are qualitatively related to it. It is an easy and most common test. The tests are conducted on cubical or cylindrical specimens. The cube specimen is of size 15 × 15 × 15 cm and the cylinder is about 15 cm diameter and 30 cm long.

The largest nominal size of the aggregates does not exceed 20 mm. The moulds are to be of metal moulds, preferably of steel or cast iron. The moulds are made in such a way that the specimen is taken out without damage. A tamping steel bar of 16 mm diameter 0.6 m long with a bullet end is used for compacting.

The test cube specimens are made as soon as practicable. The concrete is filled into the mould in 5 m deep approximately. Each layer is compacted by tamping rod (25 to 35 strokes depending on 10 or 15 cm depths) or by vibration. After the top layer has been compacted the top of the mould is leveled using a trowel. The top is covered with a glass or metal plate to prevent evaporation.

The specimens are demoulded after 24 hours and submerged in fresh, clean water or saturated lime solution and kept there until taken out just prior to testing. The water should be maintained approximately at 27° C ± 2° C and on no account the specimens are allowed to become dry. The specimen is tested in a compression testing machine at the completion of 7 days and 28 days. Compression on the cube or cylinder undergoes lateral expansion owing to the Poisson's ratio effect.

Cylindrical specimens are less affected by end restraints caused by plates and hence it is believed to give more uniform results than the cube. Further cylinder simulates the real condition in the field in respect of direction of load. Normally strength of cylindrical specimen is taken as 0.8 times the strength of cubical specimens.

2. Flexural Strength Test

Concrete is relatively strong in compression and weak in tension. Tensile stresses can develop in concrete due to drying shrinkage, rusting of steel reinforcement, temperature gradient and many other reasons. Hence, the tensile strength of concrete gains importance.

Direct measurement of tensile strength is not feasible. Beam tests are found to be dependable to measure the flexural strength property of concrete. Modulus of rupture is taken as the extreme fibre stress in bending.

The value of modulus of rupture depends on the dimension of the beam and the type of loading. The loading adopted is central or two-point loading.

In the central point loading (Figure a), the maximum fibre stress occurs below the point of loading where the bending moment is maximum.

In the two-point loading (Figure b), the critical crack may appear in any section, where the bending moment is maximum or the resistance is weak. In general, the two-point loading yields lower value of modulus of rupture than the center point loading.

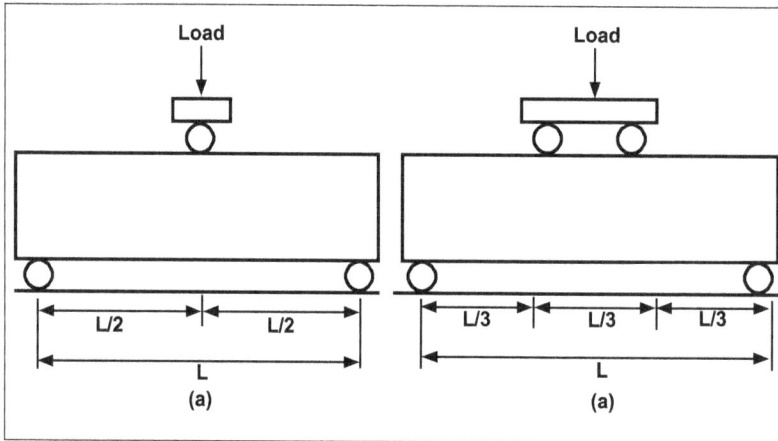

Loading arrangement in Flexural beam test.

Size of specimen is 15 × 15 × 70 cm. In case of concrete with aggregate size less than 20 mm a beam size of 10 × 10 × 50 cm may be used. Mould may be of metal or steel or cast iron. Tamping rod may be of steel of 2 kg weight, 40 cm long and should have a ramming face of 25 mm square.

Testing machine should have the sufficient loading capacity with a specific rate of loading such that the permissible errors on the applied load should not be greater than ±0.50 %. Flexural strength of the specimen is expressed as the modulus of rupture f_b as,

$$f_b = \frac{3P \times a}{b \times d^2}$$

Where,

> P = maximum load in kg applied to the specimen.

a = 17 to 20 cm for 15.0 cm specimen or > 13.3 cm for 10.0 cm specimen.

b = measured width in cm of the specimen.

d = measured depth in cm of the specimen at the point of failure.

If a is less than 170 cm for a 15.0 cm specimen or less than 11.0 cm for a 10.0 cm speci-men the results of the test be discarded.

3. Split-Tension Test

This is an indirect tension test, also referred to as Brazilian test. In this test, a cylin-drical specimen is placed horizontally between the loading surfaces in a compression testing machine. Load is applied to failure of the cylinder along the vertical diameter. The test specimen is shown in figure:

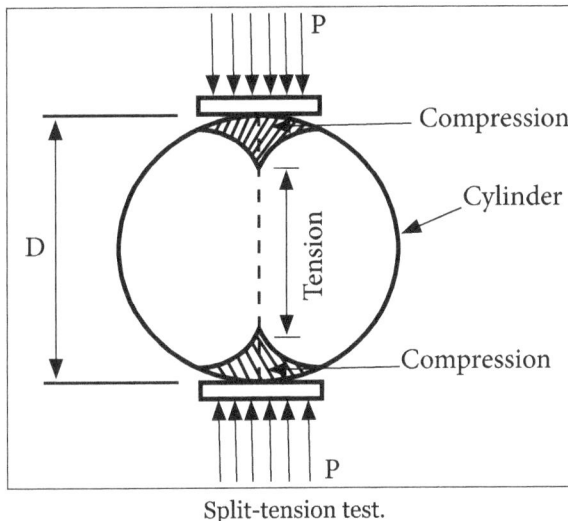

Split-tension test.

When the load is applied along the genetrix, compressive stresses develop immediately below the two generators to which the load is applied. A larger portion about 5/6th of the depth is subjected to tensile stress.

It is simple to perform and generally gives more uniform results. Tensile strength of split-tension test is almost nearer to true tensile strength than the modulus of rupture. The split-tension test gives 5 to 12 % higher value than the direct tensile strength.

Setting Times of Concrete

The hardening of concrete before its hydration is known as setting of concrete.

Following are the factors that affect the setting of concrete:

- Water Cement ratio.

- Suitable Temperature.

- Cement content.

- Type of Cement.

- Fineness of Cement.

- Relative Humidity.

- Admixtures.

- Type and amount of Aggregate.

Effect of Time and Temperature on Workability

Time and Temperature

In usual, increasing temperature will cause an increase in the rate of hydration and evaporation. Both of these effects lead to a loss of workability.

Vibration of Concrete

For proper compaction of concrete by immersion vibrators, the vibrating part of the vibrators should be completely inserted into the concrete. The action of compaction is enhanced by providing a sufficient head of concrete above the vibrating part of the vibrators. This serves to push down and subject the fresh concrete to confinement within the zone of vibrating action.

Concrete Vibration Benefits

A good vibrated concrete will offer some benefits such as:

- Have a higher compressive strength.

- A properly vibrated concrete will increase the bonding capacity between concrete and rebar.

- Provide a better sealed concrete surface reducing its permeability.

- Reduce cold joints, honeycombing and segregation.

- When the builder knows how to vibrate concrete, it can order drier mixtures that require less cement.

- Offers greater durability.

- Bonding strength between layers of concrete will increase.

- Horizontally spread layers of 20" thick, will provide with best results in concrete.

Factors influencing strength test results:

Compressive strength is the most important property of concrete. The compressive strength of concrete is calculated in the laboratory in controlled conditions. On the basis of this test result we judge the quality of concrete. However sometimes the strength test results vary so widely that it becomes difficult to reach at any conclusion. This variation in test results will be avoided by taking necessary steps.

There are 5 factors that influence strength of concrete when tested for compressive strength. These factors are mentioned below:

- Shape & Size of Test Specimens.
- Height to Diameter Ratio.
- Rate of Application of Load.
- Moisture Content in the Test Specimen.
- Material Used for Capping.

Segregation

It refers to a separation of the components of fresh concrete, resulting in a non-uniform mix. This can be seen as a separation of coarse aggregate from the mortar, caused from either the settling of heavy aggregate to the bottom or the separation of the aggregate from the mix due to improper placement. Some factors that increase segregation are:

- Larger maximum particle size (25mm) and proportion of the larger particles.
- High specific gravity of coarse aggregate.
- Decrease in the amount of fine particles.
- Particle shape and texture.
- Water/cement ratio.

Good handling and placement techniques are most important in prevention of segregation.

Bleeding

It is defined as the appearance of water on the surface of concrete after it has consolidated but before it is set. Since mixing water is the lightest component of the concrete, this is a special form of segregation. Bleeding is generally the result of aggregates settling into the mix and releasing their mixing water. Some bleeding is normal for good concrete.

However, if bleeding becomes too localized, channels will form resulting in "craters". The upper layers will become too rich in cement with a high w/c ratio causing a weak, porous structure. Salt may crystalize on the surface which will affect bonding with additional lifts of concrete. This formation should always be removed by brushing and washing the surface. Also, water pockets may form under large aggregates and reinforcing bars reducing the bond. Bleeding may be reduced by:

- Increasing cement fineness.

- Increasing the rate of hydration.

- Using air-entraining admixtures.

- Reducing the water content.

1.8 Quality of Concrete

Quality control implies that the assigned work is done according to the specifications agreed in the contract. Major civil engineering works such as multi-storeyed buildings, dams, harbors, etc. have to be constructed with at most care as they have to be in use for decades.

Specifications of work should be framed based on Government or standard procedure such that they have to serve effectively as a guide to complete the work with high quality. The specifications are as important as the design of the project. In order to make a quality concrete construction at a site the field work has to be organized with the three divisions with mutual coordination, viz., engineering division, manufacturing division and placing division.

Engineering division looks after all forms, reinforcement details and installations of all embedded parts. Manufacturing division takes care of the control of materials, batching and mixing of concrete. Placing division takes care of placing during and other subsequent works. Whole aim is to produce a quality concrete of high order and with high economics.

Requirements for high quality concrete:

- Air bubbles should be completely removed from the concrete.

- Compaction of concrete should be such that a minimum void is present.

- Adequate curing for 28 days has to be effected.

Advantages of carefully constructed concrete:

- Failure possibility is minimized.

- Lower cost of construction with long life.

- Low maintenance cost.

1.9 Extreme Weather Concreting

According to the specifications, the concrete should be placed between 15° C to 25° C. Hot weather (more than 50° C) would accelerate the rate of hardening and result in pre-matured gain of strength despite the ultimate strength remains the same. On the other hand, cold weather retards the rate of hardening concrete, although the ultimate strength remains the same.

Concreting in Hot Weather

- The amount of water needed for obtaining a certain workability increase, requiring a high water - cement ratio and/or higher cement content.

- The workability of fresh concrete decreases more rapidly. This may result in poor workability and therefore poorer compaction or it may necessitate the addition of water and therefore an increase in water - cement ratio. Both these consequences adversely affect the durability of the concrete.

- The probability of cracking of the concrete, from the surface down, is due to plastic shrinkage. This risk is greatly increased if concreting is done in hot and windy conditions. The temperature of the concrete mix and of the ambient atmosphere, the relative humidity and the wind velocity are of major importance in connection with this.

- Accentuation of thermal gradients due to additional heat released from hydration of cement. This will increase the temperature gradient in the young concrete, with high risk of cracking in consequence.

Measures required countering act the above-mentioned effects are:

- Using water-reducing admixtures. This problem can be somewhat eased by the use of retarders as well. It is advisable to add these admixtures only a short time before placing the concrete in the formwork.

- Keeping the temperature of fresh concrete as low as possible. Many specifications for works to be carried out in the hot countries state that the temperature of the fresh concrete should not exceed about 30° C. Artificial cooling will be necessary. This can be most effectively achieved by adding crushed ice to the water for mixing and fog spraying the coarse aggregates, forms and reinforcement (which then cool down quickly).

- Protection from direct solar radiation and from wind, e.g., by covering the concrete, installing screens, windbreakers, etc.

- From work to receive concrete should be wetted so as to prevent the loss of water from concrete due to absorption.

Concreting in Cold Weather

The 28 days-strength will be about 50% if cured at 0° C and of 30% if cured at -10°C. This delay in setting of concrete results in increase of labour cost, material cost, waiting time, etc.

It is the condition that freshly laid concrete under no condition should be below 4° C. In practical, the freshly placed concrete should be maintained at a temperature of not less than 21° C for 3 days or 10° C for 5 days after it is placed.

If the work has to be done below 4° C and the materials are not frozen, the heat of hydration may be increased by the use of 20 to 25% extra cement. Use of rapid hardening or high alumina cement is usually recommended.

1.10 Ready Mix Concrete

RMC is the concrete which is made at a plant away from the construction site and conveyed in special vehicles use transit mixers. It is popular equipment for transporting concrete over a long distance. Transit mixers are truck mounted having a capacity of 4 to 7 m³.

There are two variations in the ready mix concrete. In one type, the mixed concrete is transported to the site by keeping it agitated all along at a speed varying between 2 and 6 revolutions per minute.

In the other type, the concrete is batched at the central batching plant. Mixing is done in the truck mixer either in transit or immediately prior to discharging the concrete at the site.

Transit mixing permits larger haul and are less vulnerable in case of delay. The speed of the rotating drums in the truck mixer is between 4 and 16 revolutions per minute.

Here, the water needs not to be added till the mixing is commenced. With the development of a twin fin process mixer, the transit mixers have become very efficient at mixing.

Advantages of Ready-Mix Concrete:

- As the RMC companies have a laboratory facility with sufficient equipment, only quality raw materials are used to produce concrete.

- Uniform and consistent quality of concrete is assured as it is produced by automated batching plants.

- Control of the water-cement ratio is maintained as it is monitored and operated by automated batching plant.

- Due to high workable and cohesive mix, less chance of segregation, lumping and consequent absence of honey- comb in the concrete.

- Better finishing is feasible due to pump able concrete mix.

Disadvantages of Ready-mix Concrete:

- Time and rate of pouring of concrete depend on the traffic condition of the site.

- Concrete produced for a stipulated work cannot be used during unexpected situations like traffic, rain, machine problem, etc.

- Better planning is required as the concrete cannot be taken as and when required.

- There is a possibility of change of water-cement ratio by the workmen.

Comparison of Site-mix Concrete with Ready-mix Concrete

Item	Site -Mix Concrete	Ready-mix Concrete
A. Production:		
Raw material	Required Near Mixer	At Batching Plant
Weighing	Manual	Computerised
Moisture Adjustment	Approximate	Automatic
Water-cement ratio	Manual (Approx.)	Computerised
Dozing of Admixture	Manual (Approx.)	Computerised
Mixing	Tilting Mixer	Pan Mixer /Single shaft
Mixing Time	Only Approximate	Auto Timer
Batch size	0.14 Cum	0.5 Cum. to 6 Cum
Discharge	Platform	Transit Mixer / Pump
Rate of concreting	3 Cum. /Hour/ Mixer	30 Cum to 240 Cum/ Hour/ Plant
B. Quality		
Degree of Control	Fair	Excellent
Yield	Variable	Consistent
Testing Fresh Concrete	Once in 25 batch	Every batch
C. Raw Materials		
Selection / Sourcing	Client / Contractor	RMC Expert
Testing	Client / Contractor	RMC Expert
Storage	Multi location	Single point
Space requirement	Large	Limited area
D. Delivery	Manual	Transit Mixer
E. Placing		
Mode	Manual / Hoist / Crane	Concrete Pump
Rate	3 Cum./ Hour	As per requirement

1.11 Non-Destructive Testing

This test is done on hardened concrete. Here some properties of concrete are used to estimate strength, durability, elastic parameters, crack depth, micro-cracks and progressive deterioration of concrete. Using one or more of the above properties, various non-destructive methods have been developed. Some of the important methods in use are explained below.

Rebound Hammer Test

Commonly adapted equipment for measuring the surface hardness. This type of hammer consists of a spring controlled hammer mass that slides on a guide within a tubular housing. When the plunger at the tip of the hammer mass is pressed against the surface of concrete, it retracts against the force of the spring.

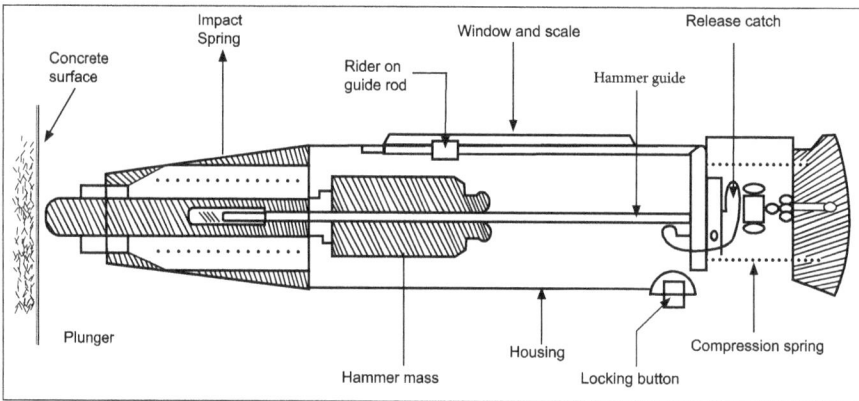

Rebound hammer.

When completely retracted the spring controlled mass rebounds, taking a ride with it along the guide. By pushing a button, the rider can be held in position to allow readings to be taken. The readings on the scale are termed as the rebound numbers.

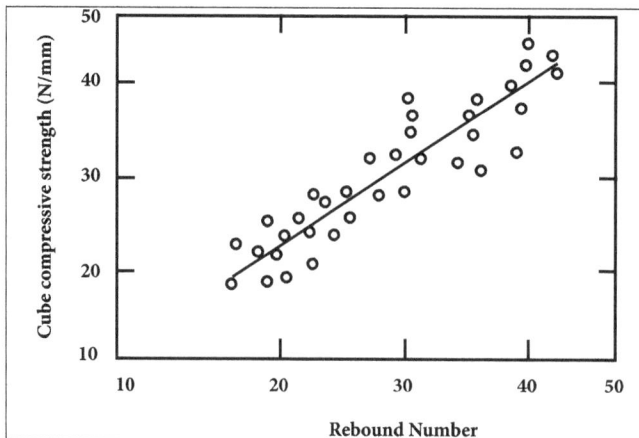

Typical rebound number and cube strength.

Calibration chart developed correlating the rebound number and compressive strength of concrete serves as a ready reference to assess the strength of in-situ concrete in members by this method.

This method provides an inexpensive, simple and quick method for nondestructive testing of:

- Concrete in the laboratory.

- In-situ concrete in precast member in precast industry.

- Concrete in members where strength is doubtful.

The test results are affected by smoothness of surface, moisture condition of test specimen, type of aggregates and the carbonation of concrete. Estimation of the strength of concrete by this method will be with an accuracy of only ± 25%.

Pulse Velocity Method

Consists of two methods, viz., Mechanical sonic pulse velocity method and Ultrasonic pulse velocity method.

Mechanical sonic pulse velocity method consists of measuring the time of travel of longitudinal or compression waves generated by a single impact hammer blow or repeated blows.

The ultrasonic pulse velocity method consists of measuring the time of travel of electronically generated mechanical pulses through the concrete. Pulse velocity methods have been used to evaluate the quality of concrete, concrete strength, durability, modulus of elasticity, the detection of water, detection of cracks, etc.

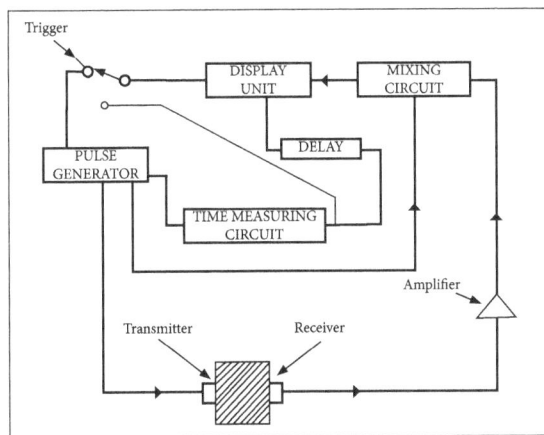

Ultrasonic pulse velocity method set up.

Out of these two, the ultrasonic pulse velocity method has gained popularity all over the world which is explained below. The ultrasonic pulse velocity method consists of measuring the time of travel of an ultrasonic pulse wave passing through the concrete to be tested.

Time of travel between the initial onset and reception of the pulse is measured electronically by the tester. The path length between the transducers divided by the time of travel gives the average velocity of wave propagation.

Pulse generator, pulse receiver and a display unit are the major units in this tester, Figure shows a typical ultrasonic pulse velocity method. Tests may be carried out in direct, indirect or semi-direct ways.

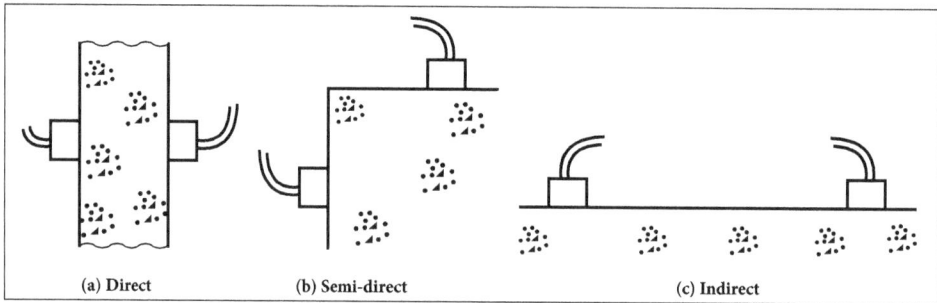

(a) Direct   (b) Semi-direct   (c) Indirect

Different test arrangements.

A calibration chart correlating the velocity of the pulse with the strength of concrete serves as reference for the assessment of in-situ concrete. High pulse velocity readings in concrete are indicative of concrete of good quality. The table gives the pulse velocity range of quality of concrete.

Pulse velocity (m/s)	General conditions
4575	Excellent
3660-4575	Good
3050-3660	Questionable
2135-3050	Poor
2135	Vert Poor

The method provides an excellent means of establishing uniformity of concrete and deserves a definite place in quality control operations. The equipment is relatively inexpensive, easy to operate and portable. The test can be carried out both on laboratory sized specimen and on large scale completed concrete structures.

Test range is up to about 3 m thickness, which can be enhanced up to 10 mm with the help of boosters. Different test arrangements are shown in Fig. Certain parameters such as wetness of concrete affect the results. The assessment of strength will be with an accuracy of only ± 20%.

Pull-out Test

Consists of pulling out from concrete a specially shaped steel insert whose enlarged end has been cast into concrete (Fig). The pull-out force is measured using a dynamometer. Because of its shape, the steel insert is pulled out along with a cone of concrete. The concrete in the pulled out region will be in shear/tension with generating lines of the cone running at approximately 450 in the direction of pull.

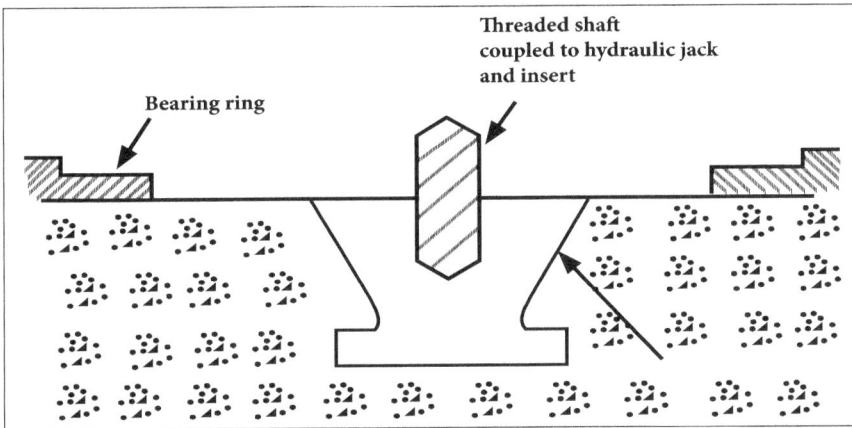
Pull - out test set-up.

The equipment is simple to assemble and operate. It is safe and testing can be carried out in the field in a matter of minutes. The tests can be reproduced with an exceptional degree of accuracy and correlation with compressive strength of concrete is good.

Drawback of this method is:

- Concrete surface gets damaged and needs to be repaired after the tests.

- The test does not measure the strength in the interior of concrete because they pull out assembly has to be inserted into the concrete at the time of concreting itself.

Frequency Method

It is a non-destructive method. It is used to determine the compressive strength and other properties. The fundamental principle on which the method based is velocity of material through a material. A mathematical relation could be made between the resonant frequency of the material to the modulus of elasticity of the material.

Break-off Method

In this method, a plastic sleeve with annular seating ring is inserted in fresh concrete to form a cylindrical test specimen and a counter bore. After hardening the sleeve is removed. Alternatively, in the hardened concrete, a concrete coring machine may be used to drill similarly shaped test specimen.

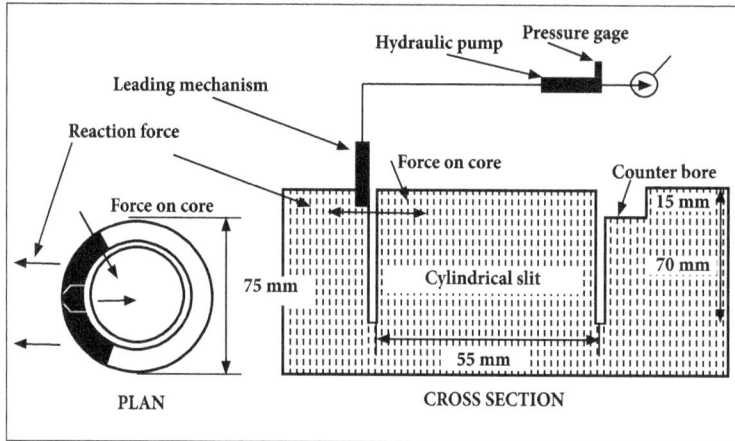

Break-off Method.

A special loading mechanism is placed on the counter bore. Hand operated pump is used to generate a force at the uppermost section of the cylinder so as to break from the concrete mass. The test result is reported as a break-off number. It is the maximum pressure recorded by the gauge measuring the hydraulic pressure in the loading mechanism. Break-off number has been correlated to compressive and flexural strength of concrete.

Codal Provisions for NDT

- The rebound hammer testing can be carried out as per IS-13311 (Pt.2).

- The pull off test is conducted as per BS-1881 part 207.

- The pull out test is conducted as per ASTM C 900-01 & BS-1881 Part 207.

- The break off test is conducted as per ASTM C 1150.

- The penetration resistance test is conducted as per ASTM C 803 / C 803-03 and BS 1881 Part 207.

2

Construction Practices

2.1   Specifications, Details and Sequence of Activities

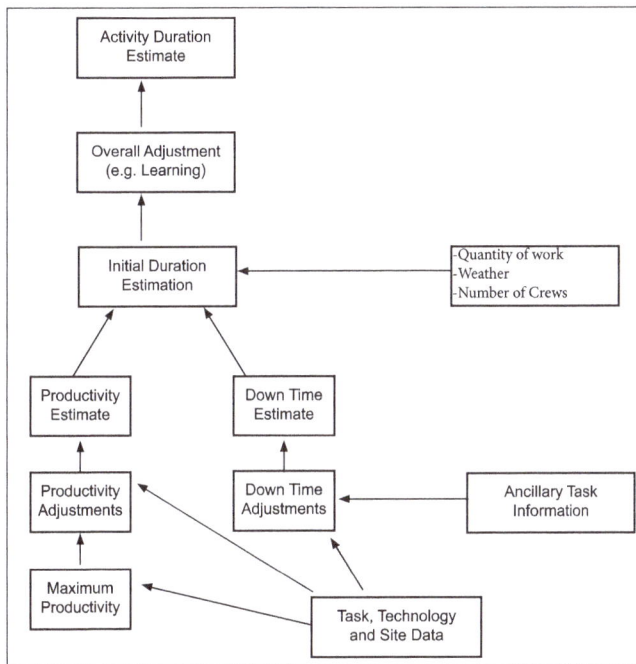

Construction sequencing is a specified work schedule that coordinates the timing of land-disturbing activities and the installation of erosion and sedimentation of control measures.

Goal of a construction sequence schedule is to reduce on-site erosion and off-site sedimentation by performing land-disturbing activities and installing erosion and sediment control practices in accordance with a planned schedule.

Construction site phasing involves disturbing only part of a site at a time to prevent erosion from dormant parts. Grading activities and construction are completed and soils are effectively stabilized on one part of the site before grading and construction commence at another part.

Key consideration of grading activities should be the coordination of cuts and fills to minimize the movement and storage of soils on, off and around the site.

This differs from the more traditional practice of construction site sequencing, in which site-disturbing activities are performed initially for all or a large section of the site, leaving portions of the disturbed site vulnerable to erosion. To be effective, construction site phasing needs to be incorporated into the overall site plan early on.

Elements to consider when phasing construction activities include the following:

- Managing runoff separately in each phase.

- Determining whether water and sewer connections and extensions can be accommodated.

- Determining the fate of already completed downhill phases.

- Providing separate construction and residential accesses to prevent conflicts between residents living in completed stages of the site and construction equipment working on later stages (USEPA, 2004).

Applicability

Construction sequencing can be used to plan earthwork and erosion and sediment control activities at sites where land disturbances might affect water quality in a receiving water body.

Siting and Design Considerations

Construction sequencing schedules should include the following:

- Design and Installation Criteria.

- The ESC practices that are to be installed.

- Principal development activities.

- Compatibility with the general contract construction schedule.

Table summarizes other important scheduling considerations in addition to those listed above.

Scheduling considerations for construction activities:

Construction activity	Schedule consideration
Construction access, entrance to site, construction routes, areas designated for equipment parking.	This is the first land-disturbing activity. As soon as construction begins, stabilize any bare areas with gravel and temporary vegetation.
Sediment traps and barriers, basin traps, sediment fences, outlet protection.	After the construction site is accessed, install principal basins. Add more traps and barriers as needed during grading.

Runoff control diversions, perimeter dikes, water bars, outlet protection.	Install key practices after installing principal sediment traps and before land grading. Install additional runoff control measures during grading.
Runoff conveyance system, stabilize stream banks, storm drains, channels, Intersystem and outlet protection, slope drains.	If necessary, stabilize stream banks as soon as possible and install a principal runoff conveyance with runoff control measures. Install the remainder of the systems after grading.
Land clearing and grading, site preparation (cutting, filling and grading, sediment traps, barriers, diversions, drains, surface roughening).	Implement major clearing and grading after installing principal sediment and key runoff-control measures and install additional control measures as grading continues. Clear borrow and disposal areas as needed and mark trees and buffer areas for preservation.
Surface stabilization, temporary and permanent seeding, mulching, sodding, riprap.	Apply temporary or permanent stabilizing measures immediately to any disturbed areas where work has been either completed or delayed.
Building construction, buildings, utilities, paving.	During construction, install any erosion and sedimentation control measures that are needed.
Landscaping and final stabilization, top soiling, trees and shrubs, permanent seeding, mulching, sodding, riprap.	This is the last construction phase. Stabilize all open areas. Including borrow and spoil areas and remove and stabilize a temporary control measures.

Limitations

Weather and other unpredictable variables might affect construction sequence schedules. The ESC plan should plainly state the proposed schedule and a protocol for making changes due to unforeseen problems.

Maintenance Considerations

Follow the construction sequence throughout the project and then modifies the written plan before any changes in construction activities are executed. Update the plan if a site inspection indicates the need for additional erosion and sediment control.

Effectiveness

Construction sequencing can be an effective tool for erosion and sediment control because it ensures that management practices are installed wherever necessary. Follow the plan and if needed, update it to maximize the effectiveness of ESC BMPs under changing conditions.

Cost Considerations

It is a low-cost measure. It requires a limited amount of a contractor's time to provide a written plan for coordinating construction activities and management practices. It might take additional time to update the sequencing plan if the current plan is not providing sufficient erosion and sediment control.

2.2 Construction Co-Ordination

Coordination is an important function in the building process. Research has shown that poor or inadequate coordination is the best that is achieved on construction sites.

A literature review carried out revealed that there is a lack of formal understanding on how day-to-day coordination is actually achieved on a construction project.

This research was directed at identifying what activities are performed to achieve coordination and which among those are the most important and more time-consuming for a construction coordinator.

Responses received from practitioners in the HongKong and Singapore construction industries indicate that identifying strategic activities and potential delays and ensuring the timeliness of all work are the most important activities.

Effectiveness of coordination methods in construction projects:

- Proper coordination is critical to the success of construction projects. Eight coordination methods used on construction projects were examined and a questionnaire was designed accordingly.

- Then 26 contractor engineers at three organizational levels working on seven large projects were surveyed and interviewed to collect information about their coordination methods.

- The analysis results indicated that coordination quality is more related to coordination effectiveness (and indirectly to project performance) than coordination quantity.

- Projects that performed well had better coordination effectiveness than projects that performed poorly.

- It was also found that engineers could spend less time on written correspondence, plans and procedures and reports without reducing their effectiveness.

2.3 Site Clearance

Reinstatements

Works should be reinstated in accordance with the specification and the contract documents. All damage done as a consequence of the works is to be reinstated to the satisfaction of the engineer and third parties involved.

Contaminated Areas

Where ground is contaminated with unwanted material, it has to be removed from the site to an approved tip and replaced with uncontaminated materials.

Disconnection of Services

The Site Representative must notify the statutory undertakers when the services to the site are no longer required.

Site Representative will be responsible for:

- Taking final meter readings and including these on the MRS sheets.

- Provide the information to the Contracts Manager for the completion of contract completion form - BAR/OPS/F/024.

Removal of Fencing

Before removing any fencing it must be established if the fence may be removed or whether it is required to be left in place until the end of the maintenance period.

Upon removing it, all post holes etc. must be filled with specified materials and all the fencing removed off site.

Removal of Advance Warning

Upon finally leaving site, all advance warning signs must be removed from site together with contract sign boards etc.

Roads and Footpaths

Prior to leaving site all roads and footpaths shall be checked for cleanliness and fitness for use by the public and where possible photographs taken.

2.4   Marking

Setting out Works

Any civil engineering work has to be set out before starting construction. Primarily excavation has to be started for which marking has to be done. Marking for excavation may be an outline marking or center line marking. Two examples of setting out of works are explained below.

Building

In order to carry out construction exactly according to plan, the outline of excavation and center line of walls are marked on the ground.

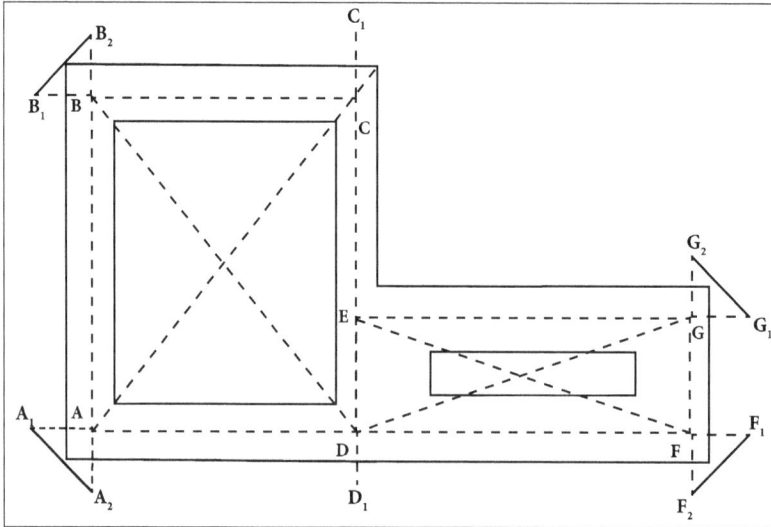

Setting out for a building.

Point	Coordinates		Remarks
	X	Y	
A1	X1	Y1	Data is computed from the left-half of the culvert
A2	X2	Y2	
A3	X3	Y3	
A4	X4	Y4	

The following procedure is:

- Based on the plan the center lines of the walls are calculated. The center lines of the rooms are set out by setting perpendiculars in the ratio 3:4:5. The corner points are identifietd as A, B, C, D, E, F and G. These points are marked by driving pegs.

- The setting of corner points are checked based on the diagonal distances AC, BD, DG and EF.

- At the time of excavation the pegs at A, B, C,... may be removed. The center lines are extended and the center points are marked 2 m away from the outer edge of excavation. Accordingly the points A_1, A_2, B_1, B_2, ... are marked with stout pegs.

- Then the width of excavation is set around the center line and marked by thread with pegs at appropriate positions.

- Further the excavation line is then marked by lime.

- Based on field conditions more pegs are driven.

2.5 Earthwork

Foundations for most structures are invariably established below the surface of the ground. They cannot be constructed until the soil or rock above the base level of the foundation has been excavated. Open excavations are supported in some soils by lateral support called bracing.

It is the engineer's duty to decide the construction procedure proposed by the builder and to check the design of the bracing. In previous soils, excavation below the water table usually requires drainage of the site either before or during construction. General aspects of excavating and providing support for the sides of the pits or cuts are discussed in the following sections.

1. Shallow Excavations with Unsupported Slopes

Shallow excavations can be made if there is enough space is available to establish slopes at which the material can stand. As a general rule construction slopes can be made as steep as possible, although a few small slides are generally not serious.

Steepness of slope depends on the type of soil or rock, climate and weather conditions, the depth of excavation and the time to which the excavation should stand.

The steepest slopes that can be used in a particular location are decided based on the experience. In sandy soils, slopes of about 1 vertical to 1½ horizontal are usually considered. Maximum slope in a clayey soil depends on the depth of cut and the shearing resistance of the clay.

2. Shallow Excavations with Sheeting and Bracing

Many a times building sites extend to the edges of property lines or are adjacent to other sites over which some structures may already be existing.

Under these conditions it is mandatory, that the sides of the excavation must be made vertical and should be usually supported.

Two common and simple methods that can be used in the above conditions are explained below.

If the depth of excavation is less than 4 m, it is common practice to drive vertical planks known as sheeting, around the boundary of the proposed excavation.

The depth of sheeting is kept near to the bottom of the excavation in progress. The depth of sheeting is held in position by means of horizontal beams called wales.

These wales in turn are commonly supported by horizontal struts extending from side to side of the excavation, Fig. The struts are usually are of timber for the excavation not more than 1.5 m wide.

For wider excavation metal pipes called trench braces are commonly used. If the excavation is too wide, the wales may be supported by inclined struts known as rakes. Rakes can be used to provide the supporting soil is firm enough to withstand the forces.

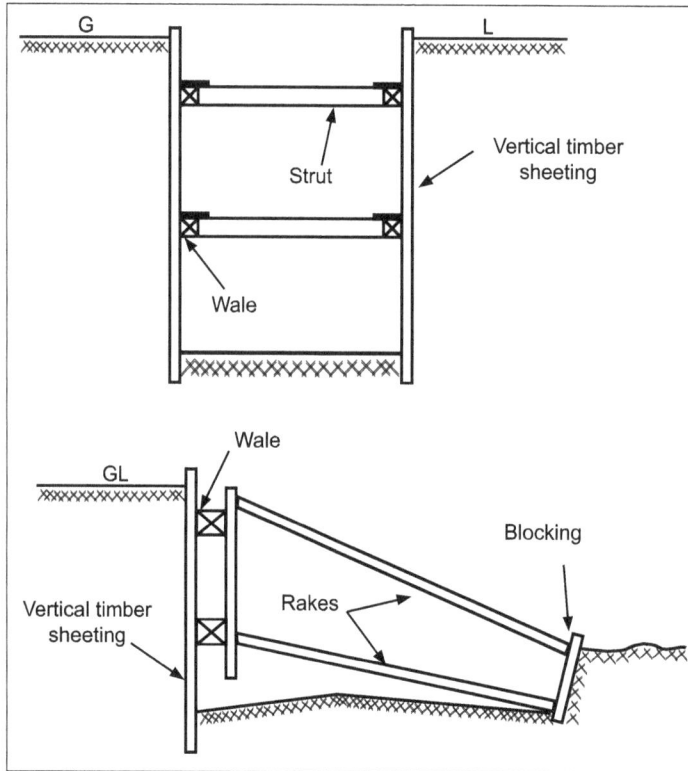

Shallow bracing.

3. Deep Excavation

Excavation beyond a depth of 1.5 m is generally categorized as deep excavation.

Problem generally encountered in deep excavations are:

- The collapsing of the sides of the trench.

- The prevention of water entering the trench from the sides or from the bottom of the trench.

Stay Bracing

This arrangement is similar to that followed for shallow excavations. This type of bracing is used in moderately firm ground and when the depth of excavation does not exceed 2 m. Here vertical sheets or poling boards are placed on opposite sides of the trench and they are held in position by one or two rows of struts.

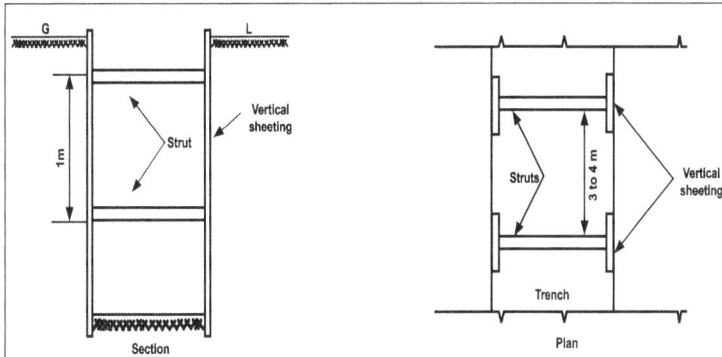

Stay bracing.

Sheets are placed at the spacing of 3 to 4 m and generally extend to the depth of trench. Thickness of poling boards are about 40 to 50 mm and of width 200 mm. Struts may be of 100 mm × 100 mm size for trench up to 2 m width and of 200 mm × 200 mm width for trench width exceeding 2 m.

Box Sheeting

This arrangement is made for loose soil and when the depth of excavation does not exceed 4 m. sheeting planks, wales and struts are used to form box like structure as shown in Figure. Here the planks are placed closer or sometimes touching each other. Longitudinal rows of wales keep the sheets in position. Struts hold the wales in position.

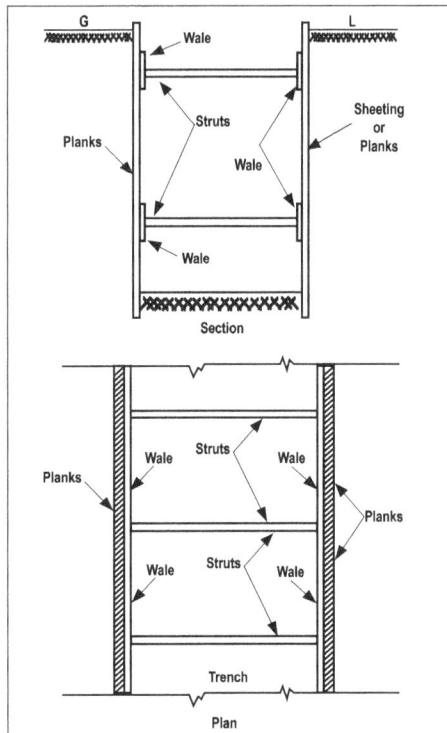

Box sheeting.

In very loose soils, additional bracing are provided. Here, the planks are placed horizontally (in plan) and are supported by wales and struts as shown in figure:

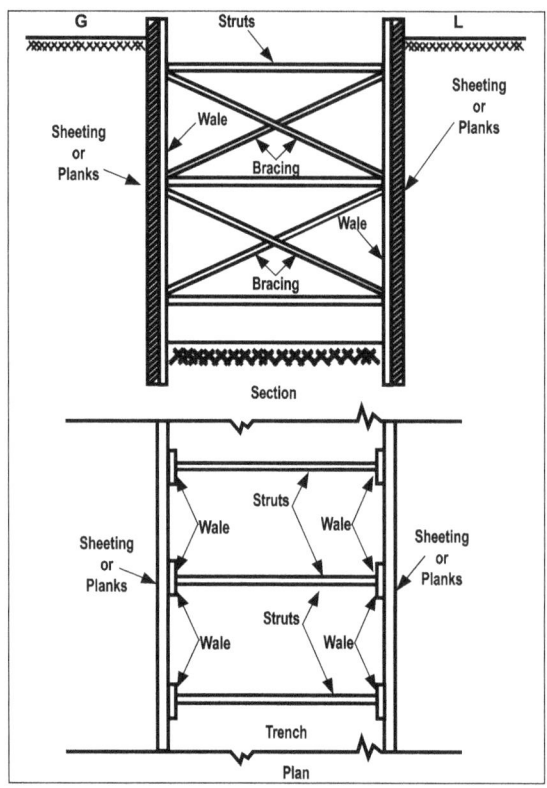

Box sheeting for very loose soils.

Vertical Sheeting

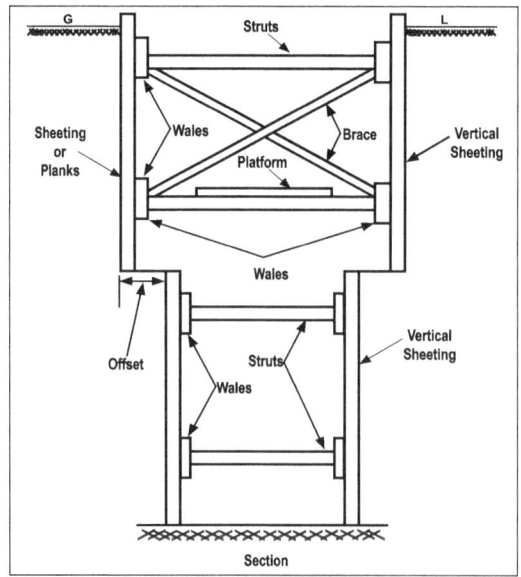

Vertical sheeting.

In soft ground, up to 10 m depth of trenches, the work is carried out in stages. This is similar to box sheeting.

At each stage of excavation one offset is provided. For each stage separate vertical sheets, horizontal wales, struts and braces are provided. Offset is provided at 3 to 4 m depth and of 50 to 60 cm wide at each stage. Suitable working platform is provided.

Runner

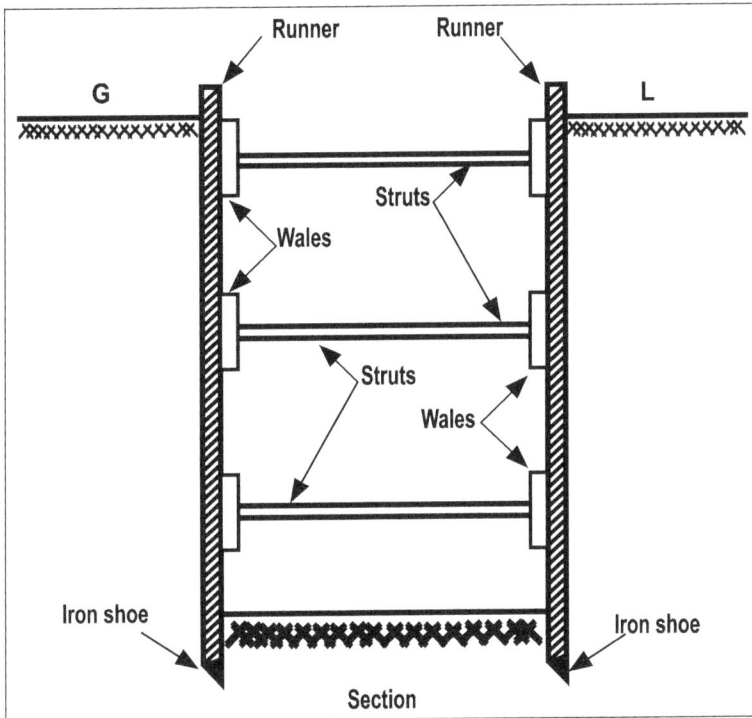

Arrangement of runners.

In situations where immediate support is needed, in case of very loose and soft ground, as the excavation progresses the special arrangement as shown in Figure is made. Here, the runners are long thick wooden sheets with iron shoe at one of its ends is used to drive the runners. The wales and struts are provided as usual.

Sheet Piling

When the depth of excavation exceeds 10 m, the use of vertical timber sheeting becomes generally uneconomical. In the above case, other methods of sheeting and bracing are commonly employed. One such procedure is driving of steel sheet piling around the boundary of the excavation. As the soil is recovered from enclosure wales and struts are inserted. Types of sheet piles commonly used are shown in figure.

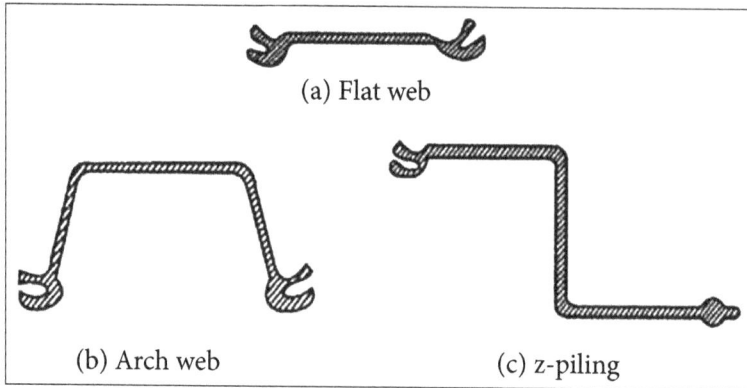

(a) Flat web

(b) Arch web   (c) z-piling

Cross-sections of sheet piling.

Strength and stiffness of piling is in the increasing order as flat, arch and z-piling. Flat and sub web types are used for shallow deep excavation whereas z-type is used for deep to very deep excavations where the heaviest pressure is expected. As the excavation proceeds wales and struts are inserted. Wales are commonly of steel and the struts may be of steel or wood.

Excavation is then preceded to a lower level and another set of wales and struts is installed. This process is continued until the excavation is completed. In order to prevent local heaves in most of the soils it is necessary to drive the sheet piles several centimeters below the bottom of excavation.

Excavation Tools

Ordinary excavations are carried out with the aid of tools such as spade, pickaxe, phaorah, crowbar, rammer, boring rod, basket, pan, etc. When rocks are encountered, excavation is carried out by using chisels, jumpers, wedges, hammers, etc. Mechanical equipment is used on large projects. Drag shovel or multi bucket trencher is used to excavate soil.

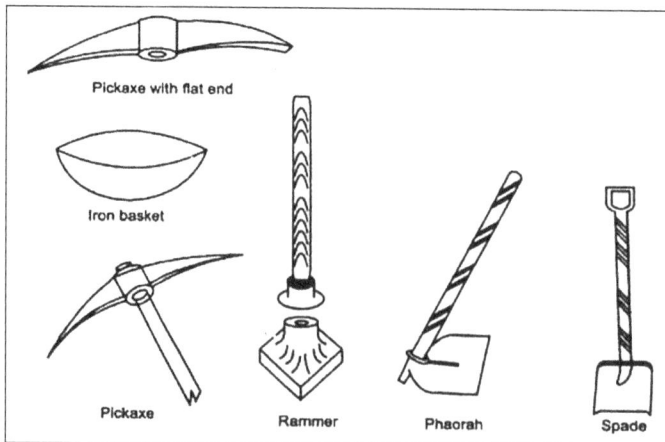

Excavation tools.

2.6   Masonry

Masonry is a structure formed by bricks, stones or hollow blocks which are arranged systematically and bonded together in mortar. It is a homogeneous mass capable of withstanding and transmitting loads without causing any failure.

Depending on the basic material used they are classified as:

- Brick masonry.

- Stone masonry or Hollow-block masonry.

Measurement of masonry works:

- Masonry is measured as "net in place" with deductions for openings such as doors, windows, ventilation etc.

- Different shapes of masonry units are measured separately such as rectangular, circular etc.

- Masonry at different heights is measured separately because masonry at higher elevation from ground may require scaffolding and hoisting.

- Masonry work is measured separately in the different categories such as:

  ◦ Facings.

  ◦ Backing to facings.

  ◦ Walls and partitions.

  ◦ Furring to walls.

  ◦ Fire protection.

- Masonry work which requires cleaning of surface is measured in area such as square meter or square feet.

- Any special treatment to masonry surfaces are measured in area such as square feet or square meter, with the count of number of coats applied.

- Any joints in masonry structure such as expansion joints or control joints are measured in length such as meter or feet with description of type of joint. Also, the type of joint filler material for the joint used is indicated in description.

- For different types of mortar or mortar with different types of admixtures are measured separately. These are measured in volume such as cubic meter or cubic feet.

- If any reinforcement is used in masonry, then it is measured separately.

- Extra items in masonry construction such as anchor bolts, sleeves, brackets and similar items that are built into masonry are described in the measurement and measured separately.

- Enumerate weep holes where they are required to be formed using plastic inserts and such like.

- Measure rigid insulation to masonry work in square feet or square meters, describing the type and thickness of material.

2.6.1 Stone Masonry

Stones are abundantly available in nature which can be properly shaped and used for construction of various parts of a building.

Similar to brick masonry, stones also can be systematically arranged to form a homogeneous mass called stone masonry. The materials used for stone masonry are stones and mortar.

Common types of stones used for stone masonry in India are: granites, sandstones, limestones, marbles and slates. Stone masonry is strong and durable.

Apart from the use in building construction, stones are used for structures like dams, piers, water-front works, monuments and memorials.

Stone masonry is not affected by dampness and does not require plastering. It will be cheap in areas where it is abundantly available.

Types of Stone Masonry

Stone masonry is classified based on the thickness of joints, continuity of courses and finish of face.

Two broad classifications are:

- Rubble Masonry.

- Ashlar Masonry.

Rubble Masonry: It is a stone work wherein blocks of stones are either undressed or roughly dressed and have wider joints.

Stones used are not of uniform size and shape but generally of pyramidal in shape to some extent. Vertical and transverse bonds have to be attained.

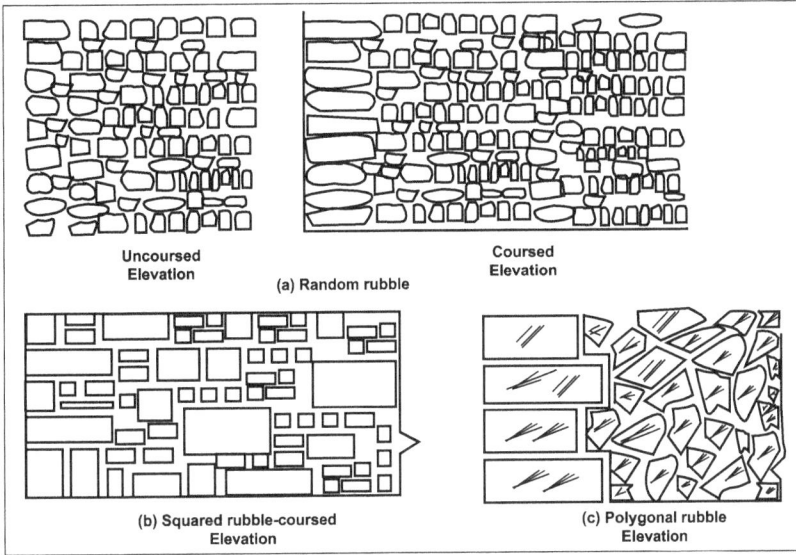

Types of rubble masonry.

Stones are provided from back to face of wall to ensure better bonding. Strength of this masonry depends on the quality of mortar, use of threw stones and filling of spaces.

There are different types of rubble masonry, viz., random rubble, squared rubble and polygonal rubble.

Ashlar Masonry: It is a stone work wherein blocks of stones are accurately dressed with very fine joints of 3 mm thick.

It is essential to ensure that the sizes of individual stones are in conformity with the general proportions of the wall.

In this masonry the beds, sides and faces are finely chisel dressed. Backing of such walls may be rubble or ashlar masonry. It is the highest grade of masonry and costly.

There are different types of ashlar masonry, viz., ashlar fine, ashlar rough tooled, ashlar rock or quarry faced, ashlar chamfered and ashlar facing. Figure shows some of the types of ashlar masonry:

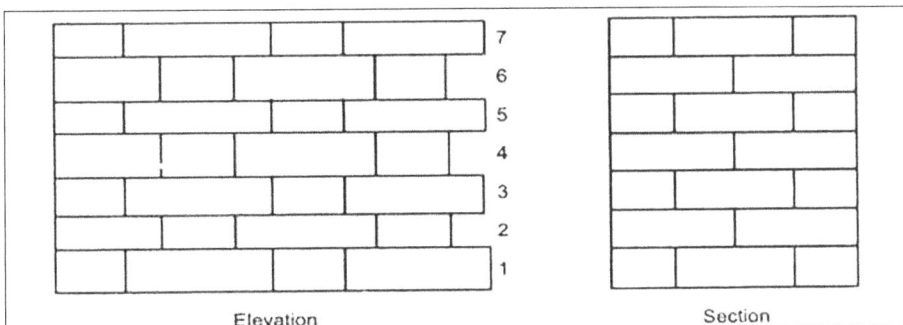

(a) Ashlar fine.

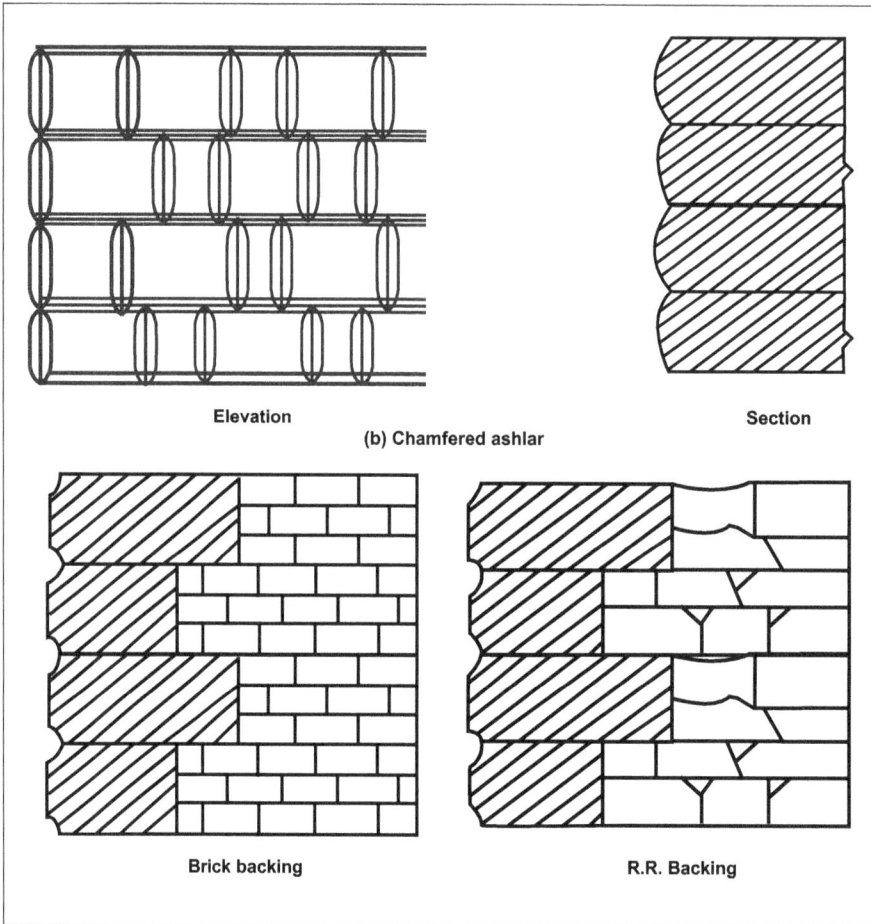

Types of ashlar masonry.

2.6.2 Bond in Masonry

Bonds in Brick Work

It is the process of arranging bricks in courses to ensure that vertical joints do not come one over the other. Wall without any continuous vertical joint shall distribute the load properly and also shall be more strong, stable and durable.

1. Stretcher Bond

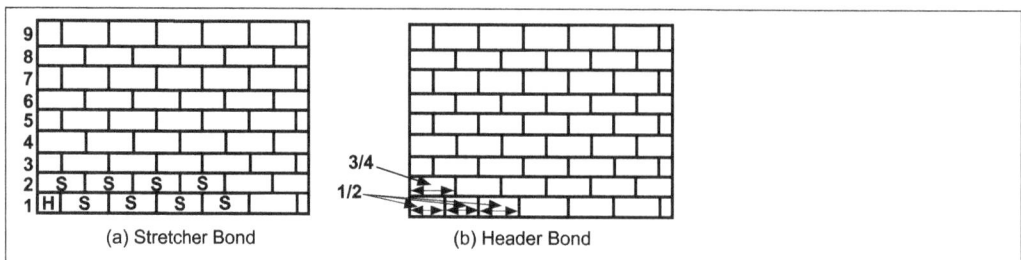

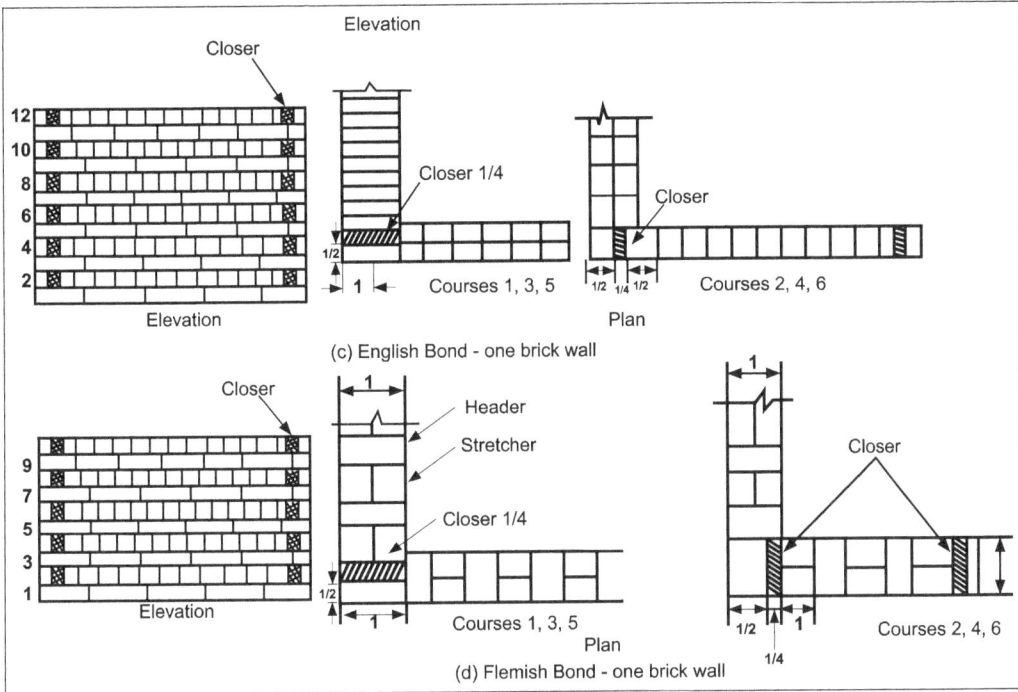

Some bonds in brickwork.

All courses are laid as stretchers. As only stretchers are visible in elevation this bond is referred to as stretcher bond. It is used for one brick and curved walls (Figure a).

2. Header Bond

All courses are laid as headers. As only headers are visible in elevation this bond is referred to as header bond. It is used for one brick and curved walls (Figure b).

3. English Bond

Produced by laying alternate courses of stretchers and headers. In order to break the joints vertically, it is essential to use a closer after the header quoin in the heading course.

Most commonly used bond which is also the strongest. Used for walls carrying heavy loads. Figure(c) shows formation of a wall adopting English bond.

4. Flemish Bond

Produced by laying alternate stretchers and headers in each. Headers and stretchers appear in the same course alternately on the front and the back faces.

Queen closer is used next to the quoin header in alternate courses in order to break the continuity of the vertical joints.

A header in any course is in the center of a stretcher in the course above or below it. Used for walls to carry moderate loads. Figure (d) shows formation of a wall adopting Flemish bond.

2.7   Concrete Hollow Block Masonry

Cement concrete hollow blocks have been in use for several masonry constructions.

1. Advantages of Hollow Block Masonry

Following advantages have given room for rapid development and use of the same in place of traditional construction materials like stones and bricks:

- Large in size but easy to handle.
- Uniformity in design.
- Easy handling and placing.
- Adequate strength.
- Attractive appearance.

2. Hollow Block Units

Two distinct types of concrete blocks are:

- Regular concrete blocks.
- Light weight concrete blocks.

Regular concrete blocks are precast cement concrete blocks made from strong aggregate and used for load bearing walls.

For partial load-bearing and non-loading bearing walls, light weight aggregates are used to make the blocks.

In general method of steam, curing is adopted which enables these blocks to be laid within a short period.

Type	Actual dimensioning cm		
	Length	Breadth	Height
Size A	39	30	19
Size B	39	20	19
Size C	39	10	19
Tolerance	± 3 mm	± 1.5 mm	± 1.5 mm

Based on the job requirement the concrete blocks may be made. There is no standard size of concrete blocks.

Concrete Association of India (CAI) recommends that the face thickness should not be less than 5 cm and the net area should be at least 55 to 60 % of the gross area.

Blocks are strengthened at the middle where cracking is liable to occur. There should be a minimum of two cores and preferably oval-shaped.

Typical concrete masonry units are shown in Figure:

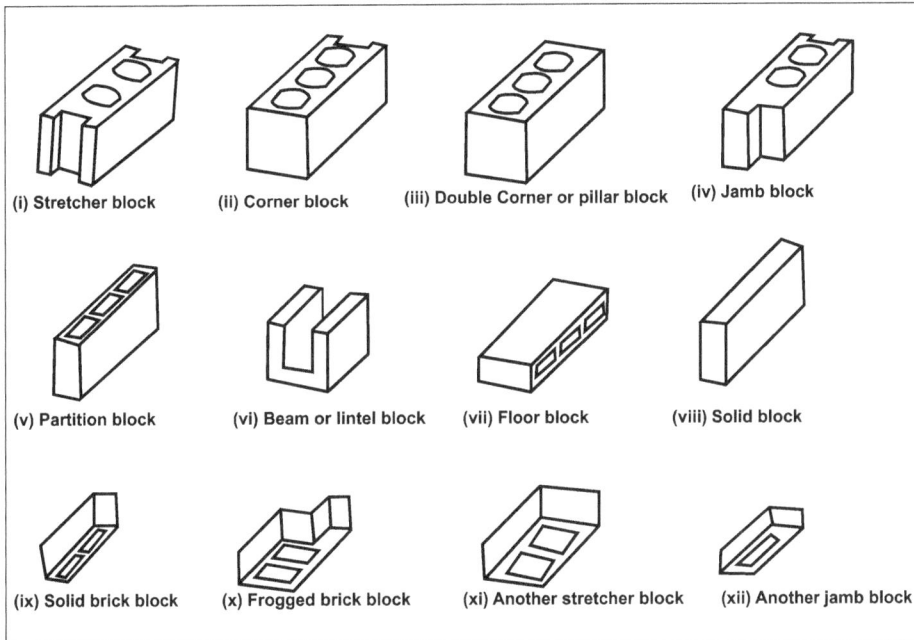

Typical concrete masonry units.

3. Laying of Concrete Hollow Block Masonry

Construction of Walls

A mortar bed is spread on the foundation concrete and leveled to have a uniform thickness everywhere. Corner block is first placed and positioned accurately.

Mortar is applied to the other end and one block is positioned to the end and aligned. Level of the course is checked after placing a few blocks.

If necessary the blocks are tapped with additional mortar such that the mortar thickness is 2 cm below and on the ends.

First course is checked to be in plumb before placing second and additional courses. As done in brick laying the successive courses are laid in such a manner so as to break the joints vertically.

For vertical joints, the mortar is applied to the projection at the sides of the block. As followed for the first course, the courses are built starting from the corners only.

Every time the verticality and horizontality are checked. All the four vertical edges of the final block and the edges of the opening are covered with mortar and pushed in position.

Face of the masonry may be pointed by running a tool. Type of joint recommended is weathered, V-shaped or concave such that the joints shed off water easily.

Construction of Columns

Columns are used whenever a large pressure to be transferred through large bearing surface. Columns may form an integral part of the wall or it can be a separate unit.

Columns are made of standard stretcher and corner blocks or other special shapes are used.

For better stability, the hollows within the blocks may be filled with plain or reinforced concrete (Figure a).

Construction of Window and Door Openings

Blocks with one hole near the opening should be filled with concrete with wooden plugs (Figure b).

Door or window frames are screwed to the wooden plugs. Fixed in the lintel with small dowels of mild steel.

Under the base of the window or door a course of solid concrete block masonry is laid which is extended into the adjacent walls up to a distance of at least 30 cm on either side.

Lintels are also of hollow channel shaped sections which can be filled with concrete and provided with steel reinforcement at their bottom.

Construction of Reinforced Walls

Construction is made by providing vertical reinforcement in the hollow of the corner block and filled with concrete. In order to increase the strength of the wall, reinforcement is provided at the horizontal joints.

Because of this provision expansion cracks which may occur due to moisture and change of temperature may be reduced. Two horizontal bars of 6 mm diameter are placed one each on the face of the wall.

Instead of steel rods welded steel mesh may be used wherever needed.

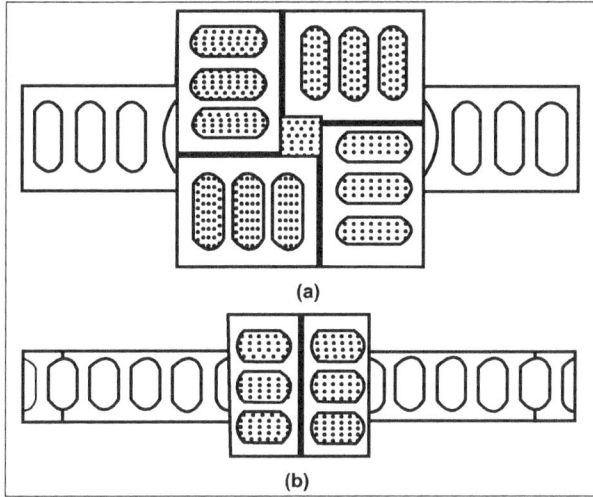

Pillasters and Piers.

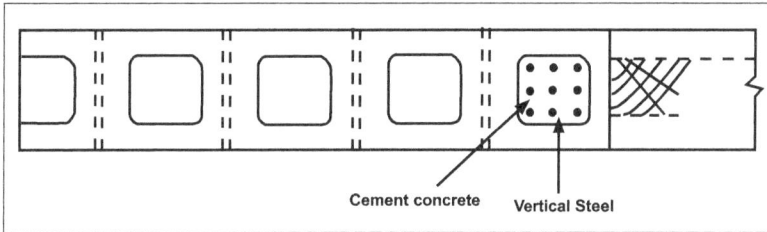

Jamb details for 200 mm thick hollow block wall.

2.8 Flooring

To create accommodation within the restricted space, floors divide a building into different levels one above the other. Bottom-most floor of a building of ground level is called a ground floor.

When the bottom-most floor is constructed below the ground level, it is called a basement floor. Floors above ground floor are top or upper floors, viz., first floor, second floor, etc.

Basic requirement of a floor is to be clean, smooth, impervious, durable and strong enough to withstand loads which come over it.

Selection of Floorings:

To select a suitable type of floor construction in a building certain factors have to be considered.

Although certain points may be common for ground, upper and basement floors some distinct different points are to be observed in each case.

1. Ground Floors

For ground floors, the selection of the type of the wearing surface is important and the other factors which need consideration are given below:

- Initial cost.
- Appearance.
- Durability.
- Cleanliness.
- Thermal insulation.
- Dampness.
- Indentation.
- Noiselessness.
- Maintenance.
- Fire resistance.

2. Upper Floors

Selection of a suitable type of construction for upper floors of a building depends on the following main factors:

- Initial cost.
- Floor loads.
- Type of construction.
- Plan of the building.
- Function of the building.
- Fire resistance.
- Sound insulation.
- Type of ceiling.
- Wearing surface.
- Weight and position of floors.

3. Basement Floor

Basement floor is not a routine type of floor provided in every building. It is provided for a particular type of buildings like apartments, hotels and restaurants, cinema halls, etc.

Selection of basement floors depends on the following factors:

- Initial cost.
- Availability of ventilation.
- Drainage of water from the floor.
- Adequate safety against fire.
- Ground water level.

Construction Practice of Ground Floors

As the ground floor rests directly on the ground, there is no need for a sub-floor. In order to drain the water outside completely adequate drainage arrangements have to be made beneath the floor.

The space above the ground, up to a height of about 25 to 30 cm below the plinth level, called as basement, is filled with some inert materials like sand, gravel, crushed stone, cinder, etc.

Over this course a damp-proof-course if needed is laid. Otherwise the floor covering is laid directly on the uniform bed.

Materials used for ground floor construction are: bricks, stones, concrete, hollow concrete blocks or wooden blocks. Materials generally used for floor coverings are: bricks, concrete, terrazzo, tiles, marbles, stones, mosaic, wood, etc.

1. Stone Flooring

Usual sizes of stones of 30 cm × 30 cm, 45 cm × 45 cm or 60 cm × 60 cm with a thickness of 2 to 4 cm are used. Square stone slabs of the above sizes are used but the slabs can be of rectangular or oblong in shape with square edges.

Stone slabs are laid on concrete bedding. Before laying the slab, a base is prepared after excavating to the required depth and the earthen base is levelled, rammed and watered.

A layer of lime concrete of thickness 10 to 15 cm is spread over which the concrete bed or subgrade is laid.

After setting the stone floor with a slope of 1 in 40, the mortar joints are raked out to a depth of 2 cm and flush pointed with cement mortar of 1 : 3.

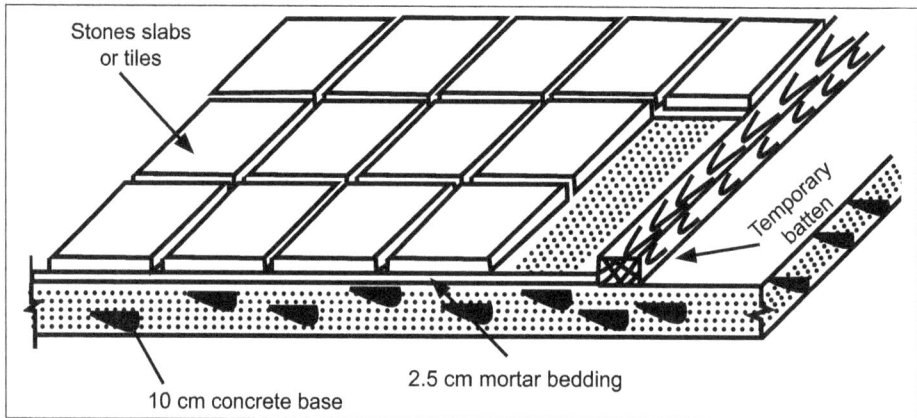

Stone flooring.

2. Brick Flooring

It is used in case of warehouses, stores and godowns. Is a cheap construction, also used in areas when stones are not available but good quality bricks are available.

It may be laid flat or on edge and arranged in herring bone pattern or at right angles to the walls.

Brick on edge is preferred compared to bricks laid flat as the brick-on-edge is less liable for crack under pressure because of the higher depth.

Bricks, in both the cases, are laid on ordinary mortar and pointed with cement or set in hydraulic mortar.

The construction of brick flooring is done as given below Figure:

- An excavation of about 40 cm depth below the intended level of the floor is made.

- The earth surface is levelled, watered and well rammed until it is dry and hard.

- Over the earth above a sub grade of 25 cm depth consisting of rubble or brick bats is laid.

- Over this a 10 to 15 cm thick layer of lime concrete or lean cement concrete (1 : 3 : 6) is laid.

- Upon this prepared subgrade, bricks are laid in the desired shape.

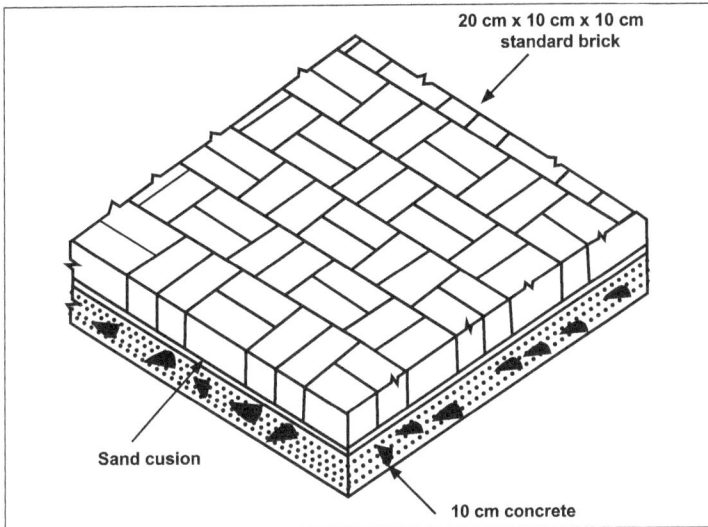

Brick flooring.

3. Concrete Flooring

It is used in all residential, commercial and public buildings. Constructed adopting either monolithic or non-monolithic construction.

In the monolithic construction, after laying the base course layer, immediately a concrete topping is provided.

In this type of construction, only a small thickness is needed for wearing surface as the bond between the base course and the wearing surface is good.

In Non-monolithic construction, the wearing surface is laid only after adequate drying of base course. Floor finish generally used is ordinary concrete finish.

Proportion of concrete finish is 1: 1½: 3. Under controlled conditions, a mix of 1: 2: 4 ratio with carefully selected aggregates may be used Fig.

For non-monolithic construction, the surface of the base concrete is brushed with a stiff broom and cleaned thoroughly. Surface is wetted and excess water removed.

Floor is laid in rectangular panels not greater than 2 m × 2 m. Alternate bays are concreted so as to avoid initial shrinkages.

When the concrete layer is even, the surface is rapidly compacted by ramming or beating and screened to a uniform level. Trowelling is done to give a level smooth surface.

Adequate curing is done for 7 days by spreading a layer of wet sand or special membrane may be used.

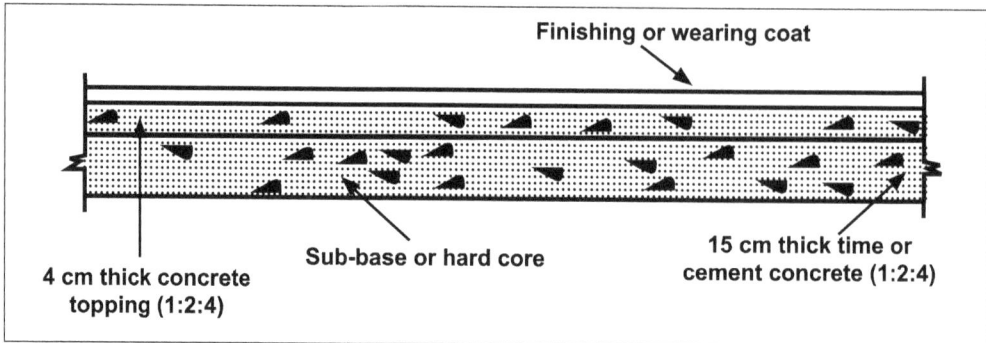

Concrete Flooring.

4. Granolithic Flooring

It is a concrete flooring with a different type of floor finish called granolithic. Granolithic finish is a concrete made of special selected aggregate.

Thickness of layer varies from 1.25 to 4 cm. If it is greater than 4 cm it may be laid monolithically or after the base concrete has hardened. It is made with very hard and tough quality aggregate in rich concrete of 1: 1: 2.

Hard fine grained granite, basalt, lime-stone and quartz stones are suitable for coarse aggregate. To get a better granolithic finish, aggregate may be crushed and used.

Fine aggregates are the natural or crushed sands with a suitable grading. Non-slippery surfaces can be obtained by adding suitable abrasives.

In case of non-monolithic construction, the base course may be prepared as done in concrete flooring. Granolithic flooring has all the advantages of concrete flooring.

5. Terrazzo Flooring

It is a special type of concrete flooring containing marble chips as aggregates. Any desired color and designs can be obtained by using marble chips of different shades and color cement.

Terrazzo mix of 1:2 or 1:3 (1 cement to 2 to 3 marble chips) is used depending on the size of marble chips. Terrazzo finish is of 10 mm thick.

The terrazzo finish is laid over the concrete base course by two methods:

*   The cement concrete base is covered uniformly by a 6 mm thick sand cushion and a tar paper is placed on this. Over this a layer of rich mortar (1:3) about 30 mm is placed uniformly.

*   A thin coat of cement is spread over the wet concrete base. This layer is cleaned and a layer of cement mortar 12 mm thick is spread evenly over it.

- When the mortar bed has hardened the terrazzo mix (1 cement: 3 marble chips) of 6 mm and 12 mm is laid after adding water and making workable mix.

- After curing for several days, the surface is polished by means of grinding machine fitted with carborundum grinding stone disc.

- During grinding, the surface is kept wet and small holes or pores are filled with a suitable cement paste matching the surface configuration.

- Surface is then washed with a weak solution of soft soap in warm water. This type is used in public buildings like banks, hotels, offices, etc., because of its decorative appearance and excellent wear-resisting properties.

6. Mosaic Flooring

In this flooring, a hard concrete base is laid first. When the base is wet a 2 cm layer of cement mortar (1:2) is evenly laid.

Over this layer, small pieces of broken tiles are arranged in different pattern. The inner space between tiles is filled with colored pieces of marble in the desired fashion.

Following this, cement or colored cement is sprinkled at the top to get a complete floor without pores. Surface is then rolled by light stone roller till an even surface is obtained.

After 24 hours of drying the surface is rubbed with a pumice stone 20 cm × 20 cm × 7 cm fitted to a long wooden handle. Polished surface is allowed to dry for two weeks before put into use.

7. Tiled Flooring

Here tiles either of clay or cement concrete, manufactured in different shapes, are used. A 15 cm thick layer of lime or cement concrete is laid over the levelled ground.

In order to receive the tiles at 25 mm thick layer of lime mortar (1:3) or cement mortar (1:1) is laid. Cement slurry is spread over the hardened mortar.

Tiles are laid flat on this bed and a cement paste is applied on the sides. Joints are rubbed with carborundum stone after allowing 2 to 3 days for setting.

Entire surface is polished with a pumice stone.

Tiled floorings are used in residential buildings, hotels, offices and other public buildings. These floors can be constructed in very short time with pleasing appearance and good durability.

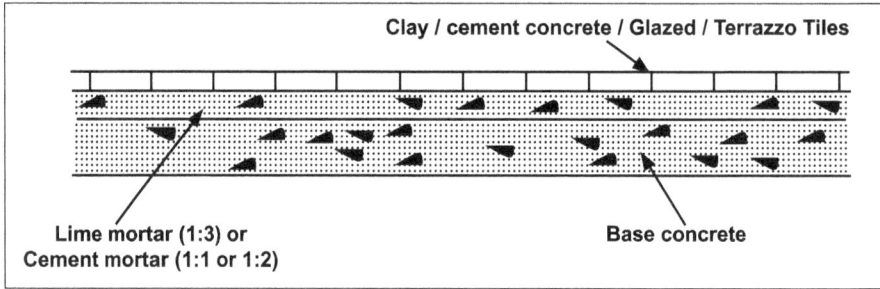

Tiled Flooring.

Special Types of Ground Floor

1. Asphalt Flooring

Asphalt floorings are of two types, viz.:

- Using asphalt tiles.
- Using mastic asphalt.

Asphalt tiles are made from asphalt, asbestos fibres and materials under pressure.

Asphalt mastic is a mixture of fine aggregates and natural or artificial asphalt.

Asphalt tiles are used to cover wooden or concrete floors. These tiles are resilient, non-absorbent, moisture proof and cheap. It is used in schools, offices and hospitals, etc.

Asphalt mastic can be mixed hot and laid in continuous sheets or pressed into blocks which can be used for flooring.

It may also be mixed with oil and asbestos and applied cold. An ordinary concrete or wood base may be used for laying this mixture.

2. Linoleum Flooring

Linoleum is a covering material generally laid over wooden or concrete floors.

Linoleum material is lubricated by mixing oxidised linseed oil with gum, resin, pigments, wood floor, cork dust and other filler materials. Available as rolls of 2 or 4 m width with 2 to 2.5 mm thick and both in plain and printed forms.

Linoleum covering are fixed to the sub-floor by means of suitable adhesive in order to have adequate bond and high durability. It is used in residential and public buildings.

3. Cork Flooring

Natural cork is the outer bark of the cork oak tree manufactured in the form of tiles and rolls like linoleum.

Cork tiles are made from high grade cork bar and are manufactured in the sizes of 10 cm x 10 cm to 30 cm x 90 cm and of 5 to 15 mm thickness. Cork tiles are available with light and heavy density.

Cork tiles are provided with joints of tongue and groove type or butt type. Rolls or carpets of cork are made by heating granules of cork with linseed oil and then compressing it by rolling on canvas. Maintenance of cork carpet is difficult.

Laying of cork flooring is similar to that linoleum covering. Cork flooring provides a warm, noiseless non-slippery, resilient flooring and with good heat insulation qualities. Preferred in libraries, churches, hospitals, broad casting studios, theaters, etc.

4. Glass Flooring

Here Structural glass in the form of tiles or blocks is used. Fitted within the frames of various types. Structural glass is available in different forms of varying thicknesses usually from 10 to 30 mm.

Frame work holding the structural blocks should be closely spaced so as to take the anticipated loads and to be safe. Used for special purposes, e.g., to transmit light from one floor to another or for a dancing hall or for a decorative purpose.

5. Plastic or EVC Flooring

Manufactured in the form of tiles. Available in various sizes and colors and shades. Laid in the similar way as cement or clay tiles. It is slippery and hence cannot be used for normal flooring.

6. Marble Flooring

Construction procedure is same as that of mosaic flooring except that marble slabs or pieces are used instead of mosaic tiles or pieces.

Preferred when sanitation and cleanliness are required as in the case of hospitals, temples, theaters and other superior type of works.

7. Timber Flooring

This type is not preferred for ground floors. If it is used as a ground floor, the prevention of dampness is most important. For fixing the timber floors on concrete slabs, longitudinal nailing strips are provided.

Planked flooring should be laid with spaces of metals spaced 1 mm apart temporarily for providing expansion joints. Strip flooring is used in thicknesses of 2 to 2.5 cm and width of 6 to 10 cm.

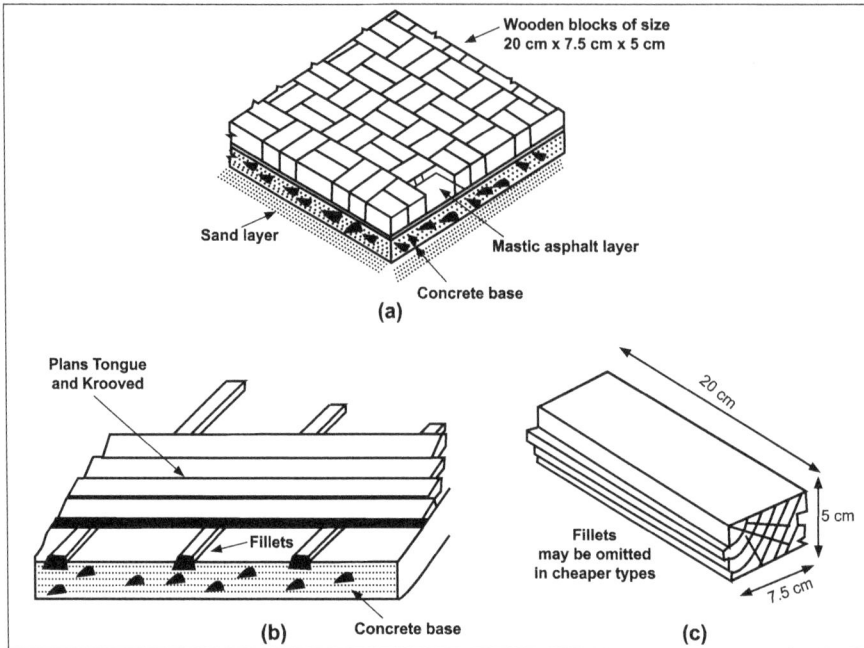

Wooden flooring details.

Construction of Upper Floors

Upper floors should be strong to take heavy loads, should have sound insulation and fire resistance and also have a good wearing resistance.

Upper floors are generally classified based on the materials of construction, arrangement of beams and girders or materials used. Commonly used upper floors are explained below:

1. RCC Slab Floor

All modern buildings are invariably constructed with reinforced cement concrete. For small spans a simple RCC slab floor is generally suitable.

For rooms, with the ratio of length of room to its width is greater than 1.5, slabs are designed to span along the shorter width. Main reinforcements are placed to the shorter width.

Thickness of the slab depends on the type of concrete used, the span, floor loads, etc. Reinforced concrete slabs are laid adopting the routine mixing, laying, finishing and curing.

Slab provides a very smooth surface at the bottom and a pleasing appearance. Accommodates all lighting arrangements.

The RCC slabs are restricted up to 4 m span beyond which beam and slab construction has to be adopted.

2. RCC Beam and Slab Floor

It is used for larger spans and heavy loading conditions. Commonly used for most of the important buildings.

Beams and slabs are designed as rectangular sections and the slabs are supported on beams. Beam used in monolithic construction is called a T-beam, i.e., a part of slab acts as a flange of the T-beam.

Main reinforcement of the slab runs parallel to the short span. In case of equal spans, two-way slabs may be constructed with reinforcement provided on both directions.

Projecting beams are covered by providing a false ceiling underneath it.

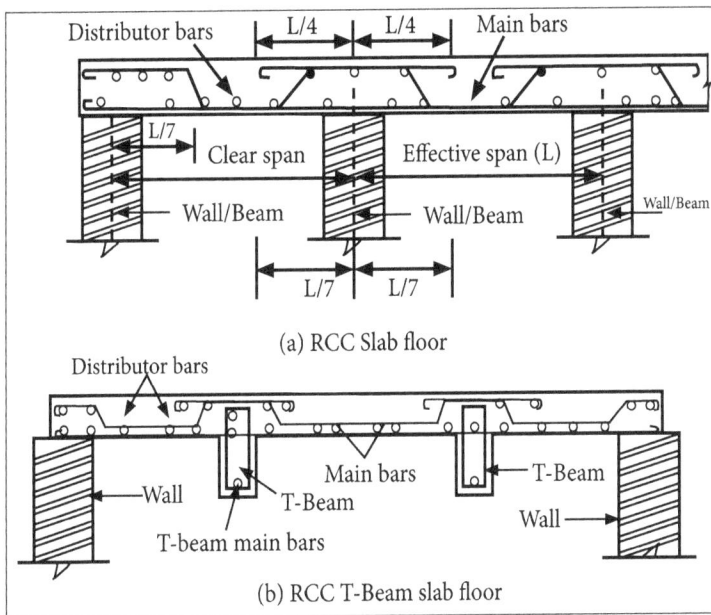

(a) RCC Slab floor

(b) RCC T-Beam slab floor

RCC floors.

The construction procedure is same as that of RCC slab floor, except for the type of centering or formwork required for the floor.

3. Flat Slab Floor

It is also called as beam less slab floor. It is directly supported on columns without any intermediate beams. Preferred where heavy loads are anticipated and where there is head room restriction.

Columns supporting the floor are invariably circular in cross-section and tops of the column are flared or tapered, which is called as capital.

Certain portion of the slab, symmetrical with the column, is thickened which is called drop panel.

2.9   Damp Proof Courses

Various techniques are adopted to prevent the defects of dampness which are explained below:

1. Damp-proofing Courses (D.P.C)

It is provided at appropriate locations for their effective use. It prevents the entry of water from ground.

In the case of a building, the best position for D.P.C is the plinth level. In case of structures without plinth, D.P.C has to be provided at least 15 cm above the ground level.

General Principles to be Adopted

In case of buildings, the following general principles should be adopted while providing D.P.C.:

- D.P.C should cover the full thickness of the walls.

- Mortar bed on which the D.P.C is laid should be level and there should not be any projection.

- In places where a vertical D.P.C is provided, it is to be laid continuously with a horizontal D.P.C and a fillet of 75 mm in radius should be provided.

- D.P.C course should be continuous and should form as a bearer from the entry of moisture.

- D.P.C should not be exposed in total.

Provisions of damp proof course in different locations at different conditions are depicted in figures:

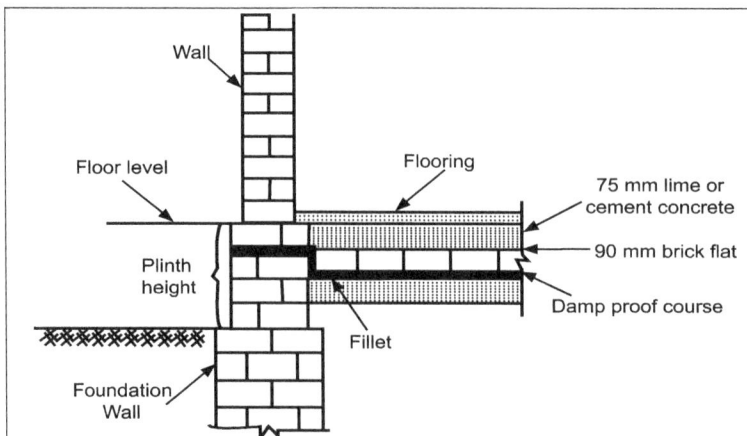

DPC above ground level for new buildings.

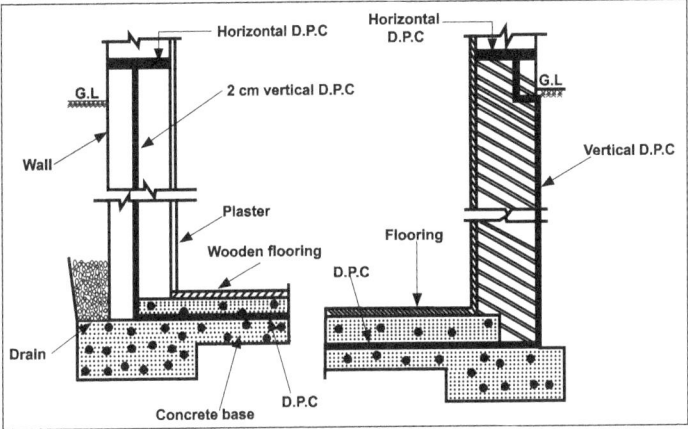

DPC in basement.

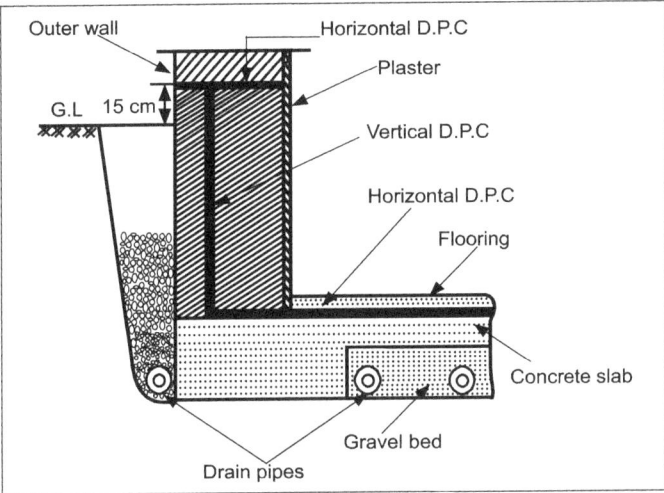

DPC treatment of foundation on weak soil.

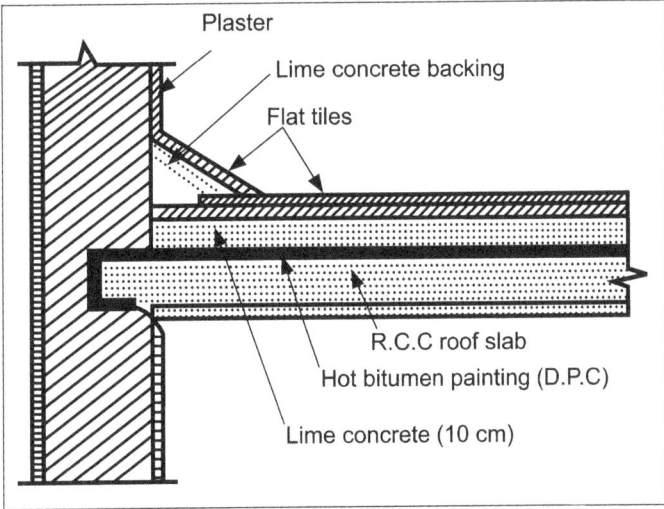

Damp-proofing of flat-roofs.

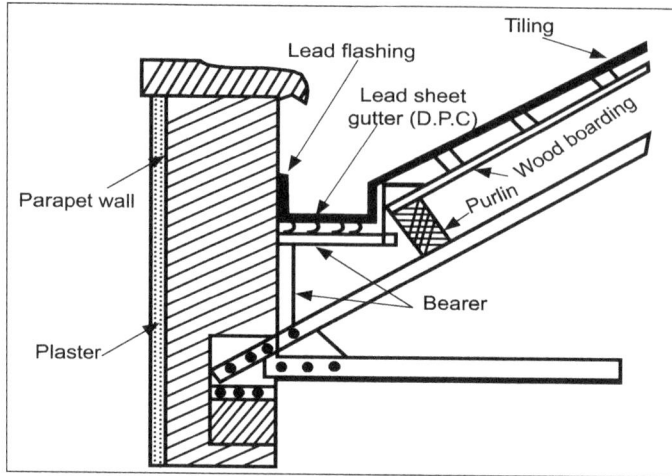

Damp-proofing of pitched roofs.

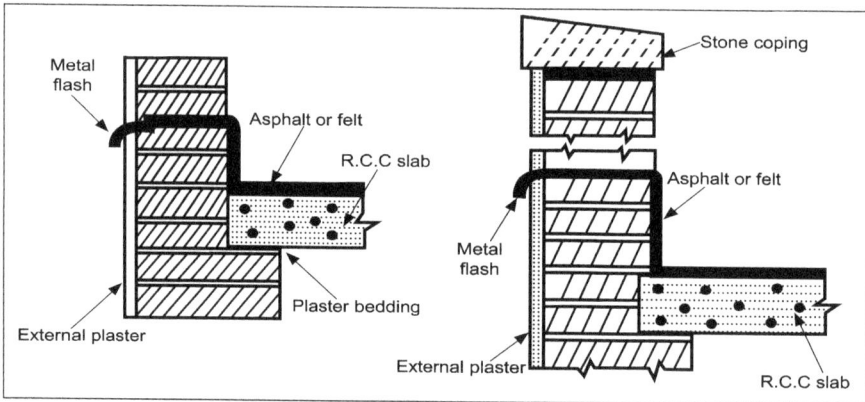

Prevention of damp along parapets.

Materials generally used for D.P.C are flexible materials like, hot bitumen, bituminous felts, bituminous sheets, polythene sheets, metal sheets of lead, copper, etc.

Semi-rigid materials like mastic asphalt or combination of materials or layers and rigid-materials like first-class bricks, stones, slates in courses and cement-concrete or mortar layers, etc.

2. Damp-proof Surface Treatment

Surface exposed to moisture is treated by providing a thin film of water-repellent material over the surface. Such a surface treatment may be external or internal.

External treatment is more effective in damp prevention compared to internal treatment. Surface treatments include pointing, plastering, painting, distempering, etc.

Lime-cement plaster mix (1 cement: 1 lime: 6 sand proportion) is more effective. Materials used for surface treatment are sodium or potassium silicates, aluminium or zinc sulphates, barium hydroxide and magnesium sulphate in alternate application, depending on the climatic conditions of the place.

3. Integral Damp-proofing Treatment

Compounds are added along with concrete or mortar while mixing which when used in construction act as barriers to moisture penetration.

Added materials function based on different principles. Based on the mechanical principle, the materials like chalk, talc, fuller's earth, etc. fill in the pores present in the concrete or mortar.

It makes the concrete or mortar denser and act as a water proofing agent. Based on the chemical reaction principle, the materials like alkalines, silicates, aluminium sulphates, calcium chloride, etc. react chemically and fill in the pores to act as water-resistant.

Based on the repulsion principle, the materials like soaps, petroleum oils, fatty acid compounds such as stearates of calcium, sodium, ammonium, etc. which when added with concrete or; mortar react with it and become water repellent.

Commercially available synthetic water repellent compounds are Pudlo, Sika, Novoid, Ironite, Dampro, per mo Rainex, etc.

4. Cavity or Hollow Wall Construction

In cavity walls, two parallel walls or leaves or skins of masonry are provided with a separation.

The components of a cavity wall are:

- The outer wall or leaf of 10 cm thick.
- An air space or cavity of about 5 to 8 cm.
- The inner wall or leaf of 10 cm thickness.

5. Guniting

Here an impervious layer of rich cement mortar (1cement: 3 sand mix) are laid for water proofing over the exposed concrete surface, pipes, etc. Rich mixture is laid by spraying using special guns.

Surface to be treated is at first thoroughly cleaned for dirt, oil or loose particles and then uniformly wetted.

A mixture of cement and sand (or fine aggregate) is then sprayed on the surface by holding the nozzle of the cement gun.

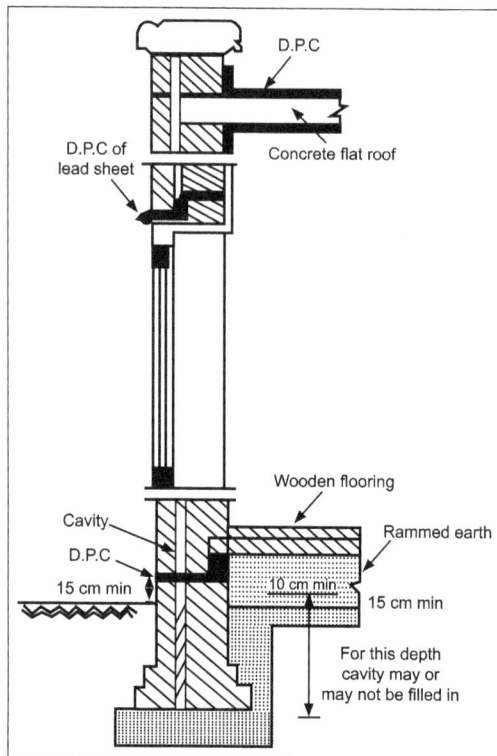

Cavity wall construction and D.P.C details for flat roofs.

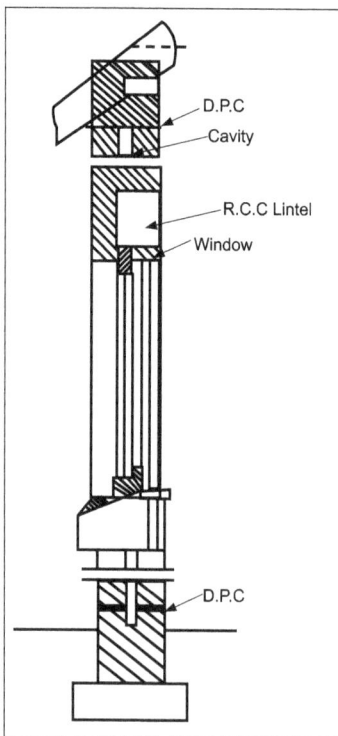

Cavity wall and D.P.C details for inclined roofs.

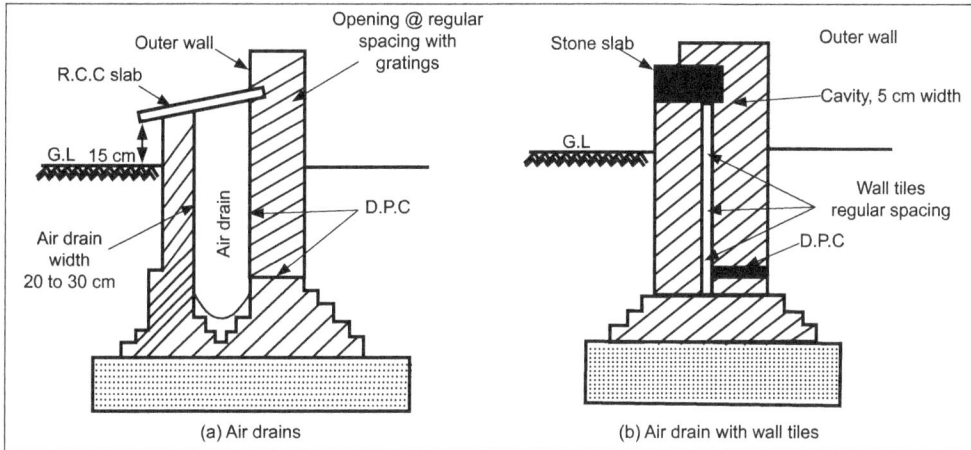

Air drains and D.P.C.

Gun is held at a distance of 75 to 90 cm from the surface and at a pressure of 2 to 3 kg/cm². After spraying the required thickness, the impervious surface is watered for at least 10 days. By this technique, an impervious layer of high compressive strength can be obtained.

6. Pressure Grouting or Cementation

It is a process by which binding or any other material in a semi-liquid form is pumped to fill the voids, fissures or cracks present in a structure. Pumped in material is called grout.

If cement grout (i.e. mixture of cement, sand and water) is pumped under pressure then the process is called cementation. More suitable for post-remedial measures for foundations and used as a major technique for ground improvement.

2.10 Construction Joints

Joints in concrete structures are provided to continue a specific work or for change in temperature.

The two types of joints usually provided in concrete structures are:

- Construction joints.
- Expansion and contraction joints.

Provided at the location where the construction is stopped at the end of day's work or for any other reason so as to bridge the old work and the new work by a proper bond. Such a situation generally occurs when large concrete work has to be executed which cannot be done within a day.

If the work is well planned such that the day's work is to be stopped at an expansion or contraction joint, in such a case there is no need for a construction joint. Construction joints may be vertical, horizontal or inclined depending on the type of structure.

In the case of inclined or curved members, the joint should be perpendicular to the axis of the structural member. Position of the construction should be well planned and constructed keeping in view the stability of the structure. Depending on the type of concrete structure the following factors should be considered:

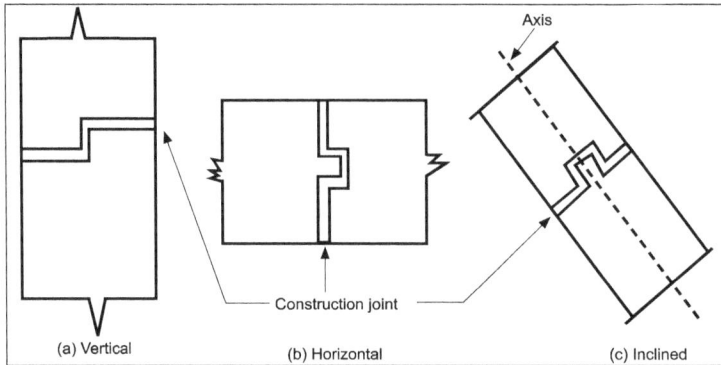

Different construction joints.

1. Columns

It should be concreted to a few centimeters below its junction with the lowest soffit of the beam. Construction above joint should be at least 4 hours after the completion of the joint.

Care should be taken such that the construction joint is at the location of least bending moment.

2. T or L Beams

Ribs of T or L beams are first concreted and then slabs forming the flanges are concreted up to the center of the rib.

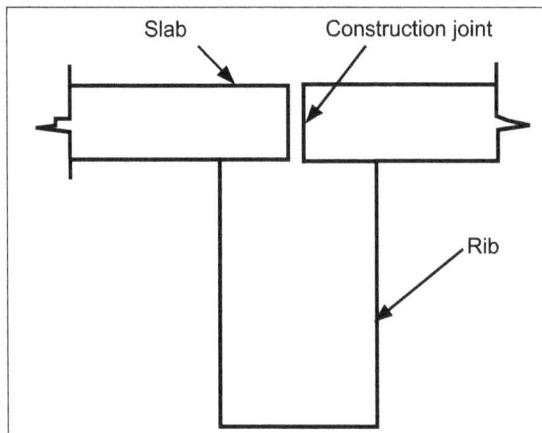

Construction joint on T-beam.

If a construction joint has to be provided between the slab and beam, the rib of the beam is concreted up to 25 mm below the level of soft fit of slab and the joint should be located at that level.

3. Simply Supported Slab

In slabs supported on two sides, the construction joint should be vertical and parallel to the main reinforcement.

Joint may be provided at the middle of the span perpendicular to the main reinforcement. For two-way slabs, the construction joint is provided near the middle of either span.

4. R.C.C Walls

Here, the location of the joint depends on the convenience in placing the framework and the access of compaction of concrete. Continuity of the joint is made by the formation of a key. Arrangement for the formation of the key and the finished position of the key is shown in figure:

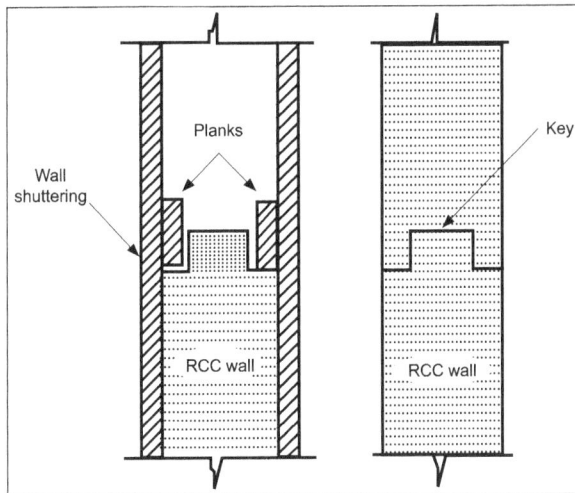

Construction joint for RCC wall.

5. Water Tanks

All water storing structures are provided with strips of copper, aluminium, galvanized iron or other corrosion resistance materials.

These provisions are called as water-stops or water-bars which are provided in a construction joint figure.

The water-stop completely seals the joint against passage of water. Natural or synthetic materials may also be used for water-stops.

Water-stops are provided in two steps. One-half of the water-stop is positioned inside the concrete and the other-half is covered during the next stage of concrete.

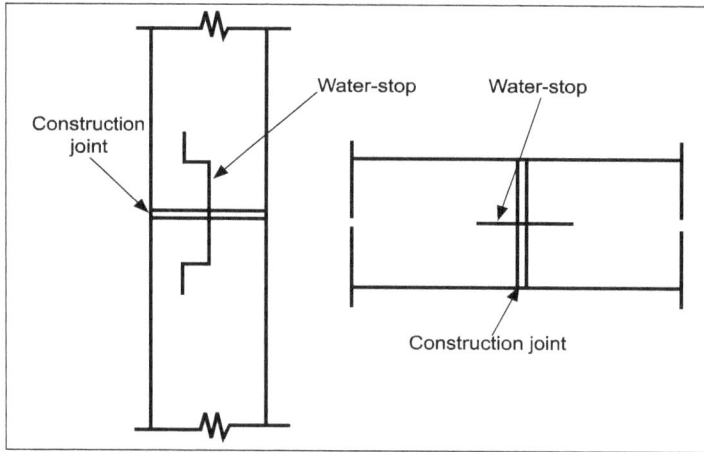

Construction joint and water-stop in water tanks.

2.11 Movement and Expansion Joint

These are provided in all concrete structures when the length exceeds 12 m. These joints are made to satisfy two requirements, viz.:

- To accommodate changes in volume of concrete due to temperature.

- To retain the appearance by maintaining the same shape of the concrete structures.

Joints are filled with some elastic materials like filter or dowels or keys. Quality of the filter should be in a position to withstand cold weather.

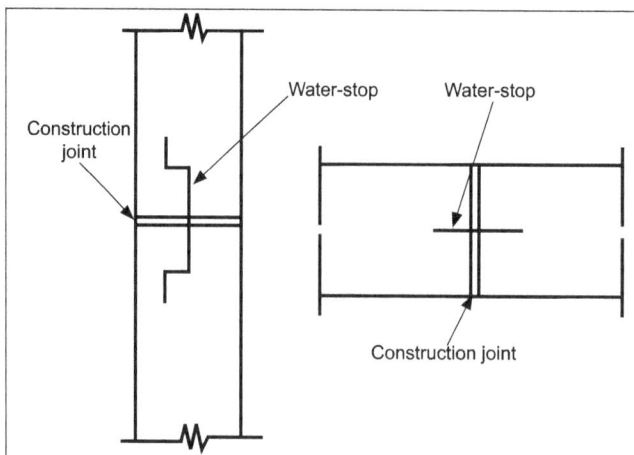

Complete contraction joint.

It should be compressible, cellular and not brittle. Conventional materials used as filter are strips of metal, bitumen-treated felt, soft wood, etc.

Provision of dowels or keys is to transfer the load. Movement due to shrinkage is controlled by the contraction joint. Contraction joint may be complete or partial.

In the complete contraction joint, there is complete discontinuity of both concrete and steel.

In the partial contraction joint, there is discontinuity of concrete but the reinforcement bars is continued across the joint.

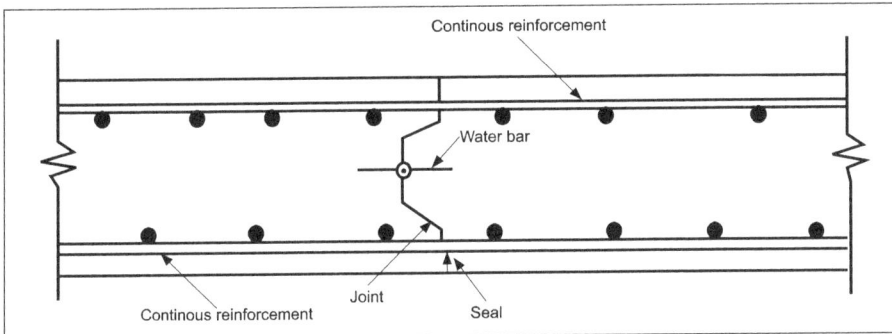

Partial contraction joint.

2.12 Precast Pavements

These are used only for light load traffic. Large size slabs may be difficult to handle.

Smaller size slabs may have more joints which may be inconvenient. Precast pavements may be used for foot paths or for lighter-road traffic.

These pavements can be used on either side of bridge as pedestrian paths. In certain cases, they may be used in wear-house flooring where heavy load is anticipated.

Before placing the precast slab as per the required pattern the subgrade and the sub-base should be properly made as detailed below.

Preparation of Grade Sub and Sub-base

Before placing pre-cast concrete slabs the subgrade and the sub-base have to comply with the following requirements:

- No soft spots should present in the subgrade or the sub-base.
- Subgrade should have adequate precision for drainage.
- Uniformly compacted subgrade or sub-base should extend 30 cm in either side of the proposed precast pavement.

- Subgrade reaction obtained through a plate bearing test should be more than 5.5 kg/cm².

- If a good or firm soil is available, no introduction of sub-base is needed.

- In general, the thickness of sub-base need not be more than 15 cm.

2.13 Building Foundations

Requirements of Good Foundation

The foundation should satisfy the following requirements:

1. Depth of Foundation

- Foundations should be carried well below the top soil, miscellaneous fill, abandoned foundation, debris or muck. If the top soils is too deep, two alternatives may be adopted.

- Top soil may be removed directly under the footing and replaced by concrete.

- Top soil in an area larger than the footing may be removed and replaced with compacted sand and gravel fill.

- Foundation should be carried below the depth of weathering.

- Foundation in sloping ground should have sufficient edge distance as protection against erosion.

- Difference in elevation of foundation should not be so great as to introduce undesirable overlapping of stresses on soil.

2. Shear Failure of Foundation

Foundation should be safe against breaking into the ground, i.e., against shear failure. To satisfy this requirement, an adequate factor of safety on the bearing capacity of the soil is provided.

3. Settlement of Foundation

Foundation should not undergo excessive total and differential settlements. Limiting total and differential settlements should satisfy the requirement specified by building codes for different structures and different soils.

Shallow and Deep Foundations

Structural foundations may be grouped under two broad categories, viz, shallow foundations and deep foundations. This classification indicates the depth of foundation installation.

A shallow foundation is one which is placed on a firm soil near the ground and beneath the lowest part of the superstructure.

A deep foundation is one which is placed on a soil that is not firm and which is considerably below the lowest part of the superstructure.

Types and Suitability of Foundations for Building

Shallow foundations are all suitable for buildings and are subdivided into a number of types according to their size, shape and general configuration.

1. Spread Footings

Most common of all types of footings with minimum cost and complexity of construction.

Provides the function of distributing the column load over a wide area taking care of the strength and deformation characteristics of the soil.

Also known as pad footings, isolated footings and square or rectangular footings (for length of footing, L and width of footing B and ratio less than 5).

2. Combined Footings

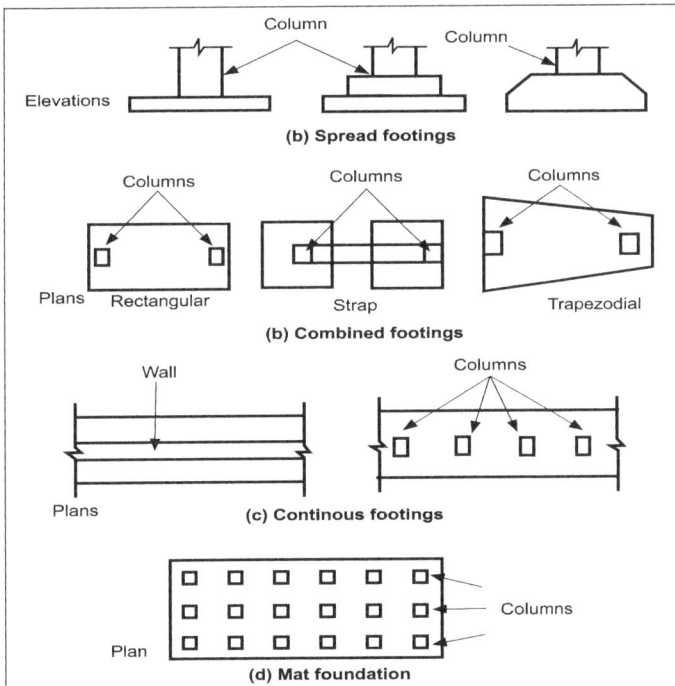

Types of shallow foundations.

These footings are formed by combining two or more equally or unequally loaded columns into one footing. This arrangement averages out and provides a more or less uniform load distribution in the supporting soil.

Distribution prevents variation of settlement along the footing. These footings are usually rectangular in shape. It may be modified to a trapezoidal shape so as to accommodate unequal column loadings or column close to property line.

It may be provided with a strap to accommodate wide column spacing or columns close to property line.

3. Continuous Footing

Carry closely spaced columns or a continuous wall such that the load distribution is uniform and load intensity is low on the supporting soil (Figure c). Also named as strip footings or wall footings (for L/B ratio greater than 5).

4. Mat or Raft Foundations

These are characterized by the feature that columns frame into the footing in two directions. Any number of columns can be accommodated with as low as four columns (Figure d).

In the majority of the cases, mat foundations are used where the soil has low bearing capacity. By combining all individual footings into one large mat, the unit pressure on the sub-soil is reduced.

Since the bearing capacity increases with increasing depth and width of the foundation and the settlement decreases with the increasing depth of foundation the advantage of mat foundation is two-fold.

Mat foundation is also preferred when the total area of the footings exceeds 50% of the total plinth area.

Pile Foundations

Piles are slender structural members normally installed by driving by hammer or by any other suitable means. Piles are usually placed in groups to provide foundation for building.

There should be a minimum of three piles. A group of piles are covered and connected together by a cap.

Reinforcement of RCC piles are continued into the cap to a distance of at least 8 cm. Cap is projected 10 cm around the piles.

Pile foundation may be considered as a column support type of foundation. Here the load of the building structure is transmitted by the piles to the hard stratum below or it is resisted by the friction developed on the sides of piles.

Based on the design load and the sub soil conditions the number of piles, length of piles, spacing of piles and type of piles are decided. Piles are installed by driving or cast in-situ.

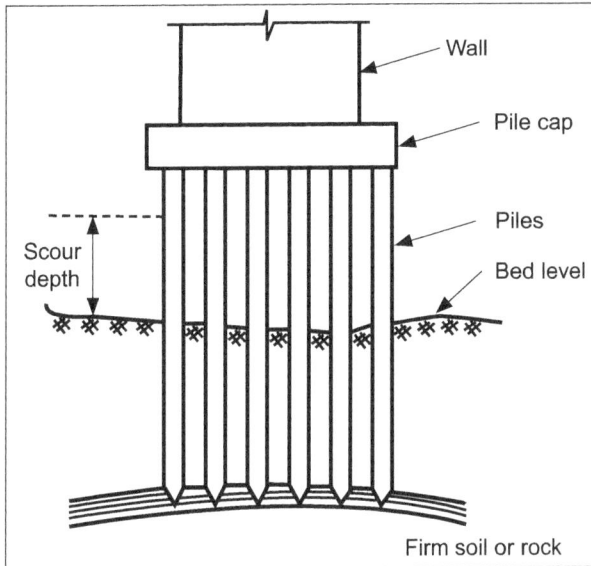

Details of a pile foundation.

Formwork for Building Foundations

The usual formwork or forms for a wall or column footing are explained below:

Formwork for wall footings:

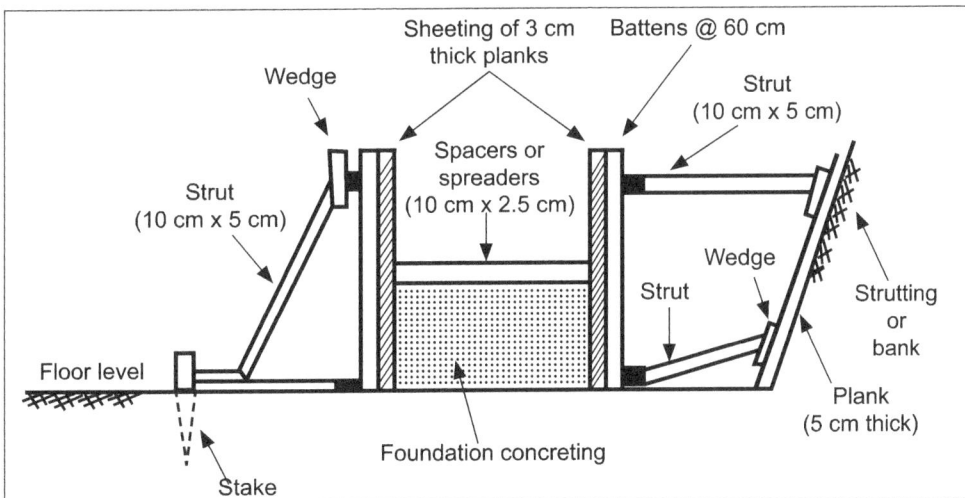

Wooden formwork for wall footing.

These formworks comprise of two planks which are easily adjusted to the width of footing by means of struts and wedges on either side.

One side of the footing is set initially to the line and positioned with the help of stakes arranged at spacing's of 2 m c/c.

Keeping this plank as reference, the other side is set with the aid of spacers at a constant distance equal to the width of the footing. After concreting reaching the position of spacers, they are removed.

Formwork of simple column footing: Based on the size, shape and nature of column footing various types of forms are used. Commonly used column footings are simple and stepped footings.

Column footings can also have sloping faces. In such cases, form faces are accordingly made sloping as per the requirement. Figure shows formwork for a square or rectangular column. Here, a box of required size is made out of planks in such a way that two ends are fixed with cleats.

Other two ends are fixed at a distance of 30 cm in excess of the actual footing dimensions. Suitable braces are provided to make the box rigid and strong.

In order to provide further safety, the opposite panels of the box are connected by tension wires which are removed along with formwork.

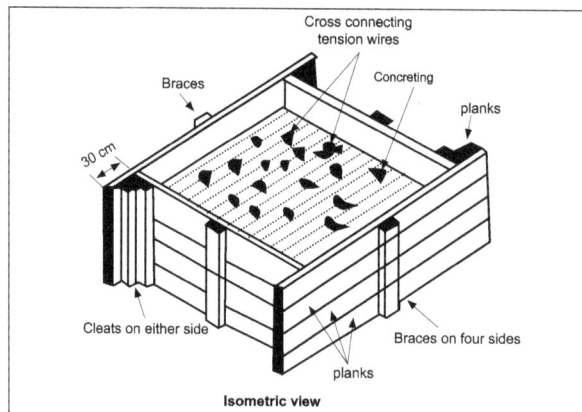

Wooden formwork for a square or rectangular column simple footing.

2.14 Basements

Construction of Basement Floor

Basement or cellar floor is the lower storey of a building fully or partly below the ground level. Basement floor is not used for residential purposes.

It is used for:

- Storage of household materials.

- Strong room for banks.

- Air-conditioning, washing machine and other service materials.

- Parking spaces.

Requirements for basement floor:

- The head-room from the basement floor should be at least 2.4 m.

- Ventilation should be adequate. This requirement is same as for ground and upper floors. Necessary provision should be made by providing air-blowers, exhaust-fans, air-conditioners, etc.

- The ceiling of the basement should be at least 0.9 m above the average surrounding ground level.

- The access to the basement shall be separate from the main.

- Alternate staircase should be provided for access and exit to the higher floors.

- The walls and floors of the basement shall be water tight.

Drainage, water proofing and damp-proofing are very important in basement. When a basement is constructed below the normal ground water level special precaution must be taken to prevent seepage into the structure.

Two general methods which are in common use are:

- Drainage.

- Water-proofing.

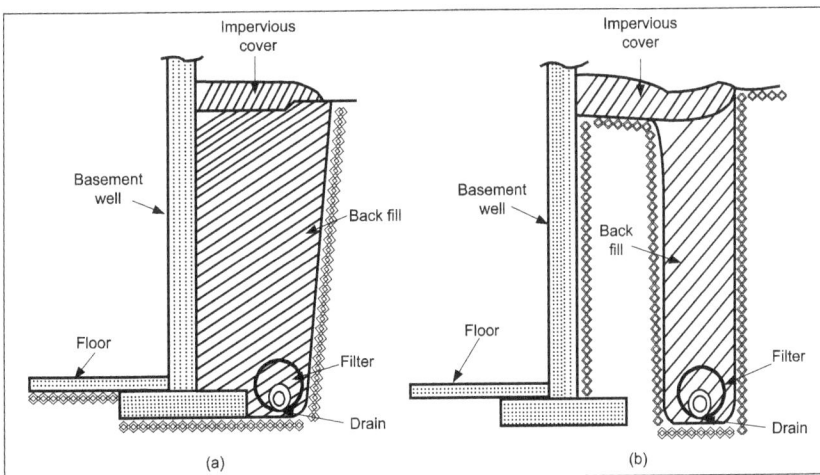

Arrangement of drains.

By adopting the second method the entrance of the water adjacent to the structure is prevented by some sort of impervious barrier.

The two methods are often combined. Drainage method may be suitable where the seepage is small enough to permit removal of the water usually by gravity flow into sewers or ditches.

Floor drains are not usually necessary if the footing drains are effective. It is advisable to provide where there is slow seepage from beneath the structure.

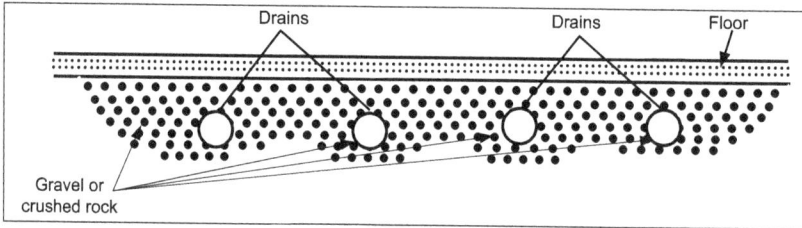

Floor drains.

If the seepage is excessive in drains, it may be necessary to water proof the basement and permit the structure to be subjected to the full water pressure.

Most positive procedure is the membrane or damp-proofing method. In this method, a membrane consisting of alternate layers of fabric and bituminous material is constructed on or near the exterior of the building. Bituminous materials are applied hot.

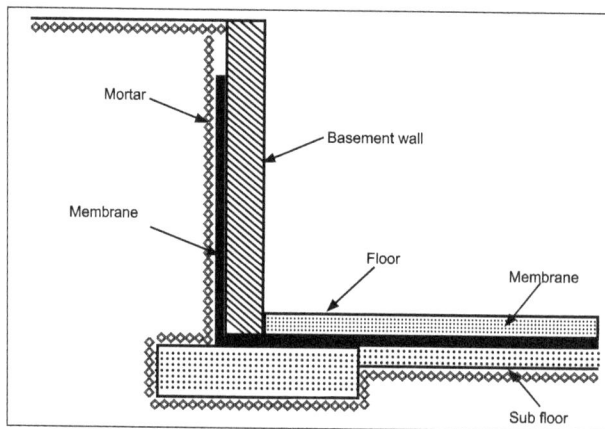

Arrangement of membrane and Sub floor.

2.15 Temporary Shed

Temporary shed construction will be the first step before we bring our material near our site. We need to construct temporary shed to keep our construction material and our watchmen will stay there to look after our site and material.

We need to construct this in our neighbor's site with their permission. So, have a plan to contact them early. Since it is temporary, we can consider using soil instead of cement.

2.16 Centering and Shuttering

Concrete is in a plastic state initially and has to be kept within an enclosure of a desired shape by proper supporting till it gains adequate strength. This temporary enclosure is known as formwork or shuttering or simply as forms.

For circular works such as arches, domes, etc. the term centering is used generally instead of formwork or shuttering.

Term moulds is used to indicate formwork of relatively small units such as lintels, cornices, cubes for testing, etc.

Requirements of Formwork

Irrespective of the type of material used the formwork should satisfy the following requirements:

- Adequate strength.
- Smooth inner surface.
- Enough rigidity.
- Quality.
- Less leakage.
- Economy.
- Easy removal.
- Supports.

Materials Used

The choice of the material to be used for the preparation of a formwork is based on the nature of the job or economy or both. The material generally used are timber and steel. Sometimes plywood and aluminium are also used.

1. Steel Formwork

If it is intended to reuse the formwork material for several times, it is preferred to use steel or aluminium. Initial cost of steel is very high. Sizes of steel sections should be decided based on the requirements.

Advantages of Steel Form Work:

- It is easy to install and dismantle and so there is less labour cost.
- ii. It is of high strength and durable.
- It can be designed precisely.

2. Timber Form Work

Timber formwork cannot be used many a times as that of steel formwork. Timber used for formwork should be well seasoned and should be neither too dry or too wet.

Components of timber formwork depend on the design load and the type of timber available. Number of nails should be less and the heads projecting out for easy removal.

Advantages of Timber Form Work:

- It is cheap and the initial cost is less.
- It can be easily altered for another work with modification.
- It is used for small works requiring less repetition.

3. Plywood Form Work

Advantages of Plywood Form Work

Now a days plywood is used as form-work for light loads. Advantages compared to timber formwork:

- It gives surfaces which are plain and smooth and may not require any further finishing treatment.
- Because of large size of plywoods are available, it is possible to cover large area and less labour cost is involved.

2.17   Slip Forms

It is also called as climbing forms. During the process of construction, the slip-forms are raised while the concrete is in a plastic state. These are used for the construction of tall structures, e.g., chimney.

Slip-forms may be classified as:

- Straight slip-forms.
- Tapering slip-forms.

- Sand slip-forms for special applications.

Straight slip-forms are used for silos, straight-chimneys, water towers, etc. Tapering chimneys and ventilators are constructed using tapering slip-forms.

Once a set of slip-forms is completely assembled at the bottom of the structure, the forms are filled slowly with concrete. Concrete is allowed to gain sufficient rigidity.

Once attaining sufficient rigidity, the upward movement of the forms is started. It is continued at a speed that is controlled by the rate at which the concrete sets.

The raising of the slip-forms is done with the help of screw jacks or hydraulic jacks. Among these two jacks, hydraulic jacks are superior to screw jacks and they assure uniform upward movement.

Rate of lifting of form will depend on the temperature of concrete which controls the rate of set. Concrete will stick to the forms if the lifting is slow and it will be difficult for lifting the forms.

On the other hand, if lifting is fast the concrete will bulge at the bottom of forms. Such bulge will affect the stability of the entire structure.

Advantages of Slip-forms:

- As the forms are lifted continuously, there is no possibility of any joints.
- Because of the non-formation of construction joints, there will be perfect water tightness.
- As the construction work is continuously done, time lost in removing and re-setting of forms is saved.

Difficulties faced in the construction using Slip-forms:

- Good site organization and coordination are needed.
- There is a need for large number of equipment.
- Because of continuous operation day and night shifts are to be arranged.

2.18 Scaffoldings

In the routine construction of work, sometimes it is necessary to have some temporary structure or support to continue the work.

When the height of construction exceeds about 1.5m a temporary structure, usually of timber, is erected close to the work so as to provide a safe working platform for the workers and to provide adequate space to keep the working materials.

This temporary structure is known as scaffolding or simply a scaffold. Such temporary structures are used in construction, demolition, maintenance or repair work.

Parts of Scaffolding

One or all of the parts mentioned below are used in the construction of a scaffolding for a specific work.

Standards: These are the vertical members of the scaffold which are either supported on the ground or embedded in to the ground or on sand filled drums.

Ledgers: Horizontal members of the scaffold.

Putlogs: Transverse pieces which are placed on ledgers and perpendicular and supported on the wall.

Transoms: Putlogs whose both ends are supported on the ledgers.

Bridges: These are used to bridge an opening in a wall and support one end of the putlog at the opening.

Braces: Cross or diagonal pieces fixed on the standards.

Guard Rail: Horizontal member provided like a ledger at the working level.

Toe Board: Placed parallel to the ledgers and supported between the putlogs.

Made as a protective measure to work on the working platform.

Raker: Is an inclined support. All these members are kept in position securely by means of devices such as nails, bolts, ropes, etc.

Requirements of Scaffolding

- The method of erection should be easy with less wastage of material.
- It should possess adequate strength and stability during the entire period of usage as men have to use at all heights.
- The material required for scaffolding should be available easily in all sizes and lengths.
- It should be possible to interchange the material for other works with less wastage.
- The initial cost should be comparatively less and should have high scrap value.

Types of Scaffolding

Different types of scaffolding are explained below:

- Single scaffolding or Bricklayer's scaffolding.
- Double scaffolding or mason's scaffolding.
- Cantilever or needle scaffolding.
- Suspended scaffolding.
- Trestle scaffolding.
- Steel scaffolding.
- Patented scaffolding.

1. Single Scaffolding or Brick Layer's Scaffolding

Particularly used in the construction of brickwork. It consists of a single row of standards placed at a distance of about 1.20 m from the wall. Spacing between the standards is about 2 to 2.5 m. Ledgers are fixed at a vertical distance of 1.20 to 1.80 m on the standards.

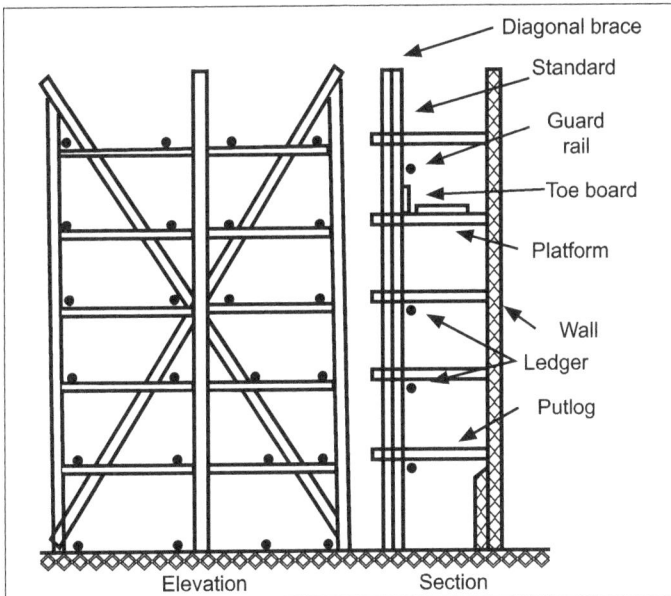

Details of single scaffolding.

The putlogs are placed at a horizontal spacing of 1.20 to 1.80 m. Also called as putlog scaffolding.

2. Double Scaffolding or Mason's Scaffolding

This is stronger than the single scaffolding. Similar to that of simple scaffolding except two rows of standards are used, out of which one is close to the wall and the other is 1.2 to 1.5 m away from the face of the wall.

No holes are made on the wall for putlogs and the putlogs are supported at both ends on ledgers. Sometimes diagonal bracing and inclined supports called racking shores are provided. It is used for stone-masonry construction.

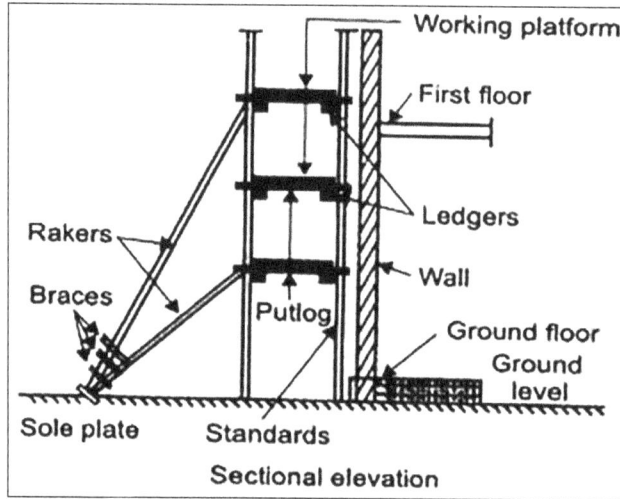

Details of double scaffolding.

3. Cantilever or Needle Scaffolding

General frame work may be of single or double type of scaffolding. Standards are supported by needles or ties.

These ties are projected out at floor levels or through openings or through holes provided in the masonry. Two types of cantilever scaffolding are shown in figure:

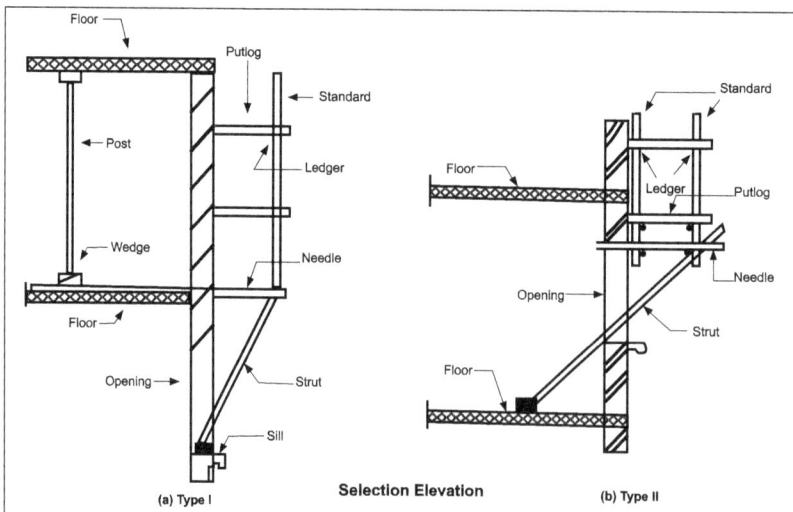

Cantilever scaffoldings.

4. Suspended Scaffolding

It is a light type of scaffolding used only for maintenance works like pointing, white washing, etc. Working platform is suspended from the roofs.

Special arrangements are made with pulleys, ropes, etc. to suspend the platform from the roof and to raise or lower based on the need.

This type of arrangement is preferred as it does not create any obstruction on the ground and only a minimum space is required.

5. Trestle Scaffolding

Here the working platforms are supported on tripods, ladders, etc. which are mounted on boggies, wheels or Lorries. It is suitable for minor repairs or painting work within a height of 5 m.

6. Steel Scaffolding

Here steel tubes are effectively used in the place of timber. 40 to 50 mm tables of 5 mm thick are used.

Tubes are commercially available in suitable lengths with special couplings and set-screws. There is no height restriction. It is strong and more durable.

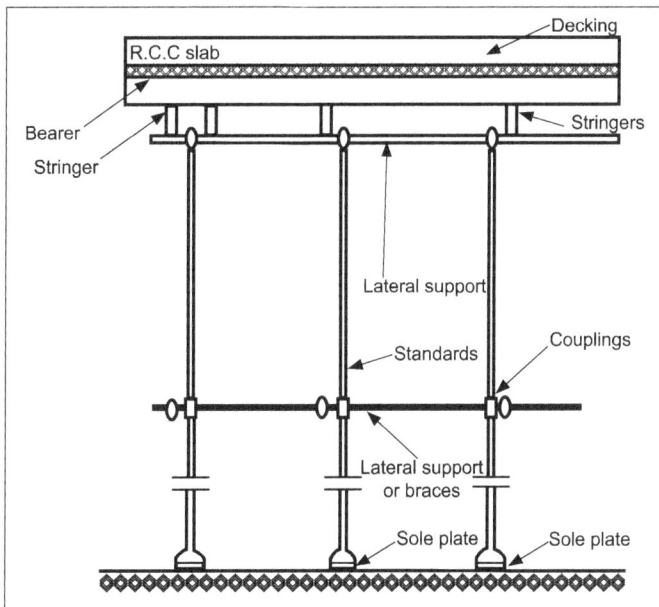

Details of tubular steel scaffolding.

Here Erection and dismantling are very easy. Material of tubes is resistant to fire and high scrap value.

7. Patented Scaffolding

It is also called as ladder scaffolding. It is a modified form of double scaffolding but of steel. These are patented scaffolds readily available in market with special types of couplings and frames.

Working platform is supported on a bracket which can be adjusted to any suitable height. It is used for light works like painting or decoration.

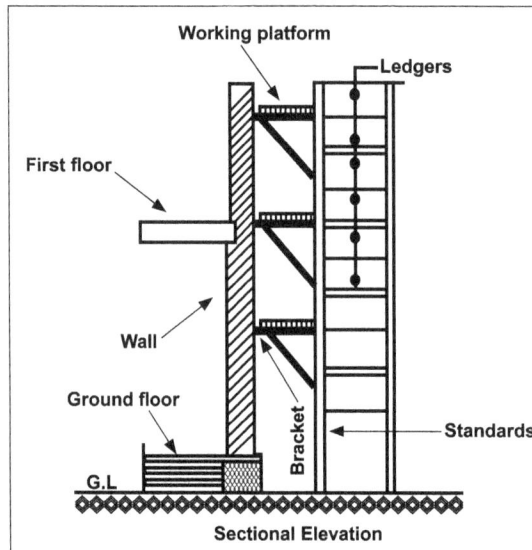

Details of patented scaffolding.

2.19 De-Shuttering Forms

As a general rule the concrete may be left in place for as long as possible up to a maximum of 28 days in normal weather conditions.

Period to which the concrete should be left in place depends on the temperature of air, the shape and position of structural member, the load condition and the type of cement.

In case use of rapid-hardening cement, lighter-temperatures, low water-cement ratio and higher loads, the curing period may be reduced and early removal of forms may be permitted.

Hardness of concrete can be checked by striking the concrete and if a metallic sound is heard it can be presumed that the concrete is hardened and the forms may be removed.

For other reasons such as re-use of forms, early use of the structure and need to cool the concrete in massive concrete the forms may be removed subjected to the condition that the concrete is adequately hardened.

In normal conditions with the atmospheric temperature is above 20°C, normal OPC is used. Type of structure, the forms may be removed after a curing period as given in table.

The forms should be slackened gradually in order to prevent the sudden imposition of load on the structure.

2.20 Fabrication and Erection of Steel Trusses

Steel Trusses

Steel roof trusses are designed in such a way that the members are either in compression or in tensions only.

Members of a truss which form the roof depends upon the span, roof slope, covering materials, center to center of trusses, etc.

Tee-section is the most suitable section for principal rafter. For struts either angle iron or channel section may be used. Round or flat sections are used for tension members.

Built up sections are also used in certain cases. Members of a truss are connected by bolts, rivets and thin plates called gussets.

1. Trusses for Small Spans

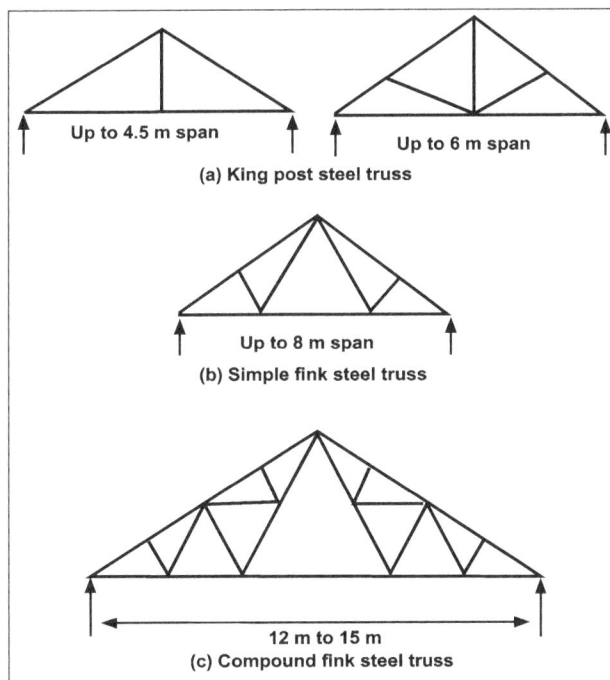

Up to 4.5 m span
Up to 6 m span
(a) King post steel truss

Up to 8 m span
(b) Simple fink steel truss

12 m to 15 m
(c) Compound fink steel truss

9 m to 12 m

(d) Howe steel truss

12 m to 15 m

(e) Compound Howe steel truss with raised chord

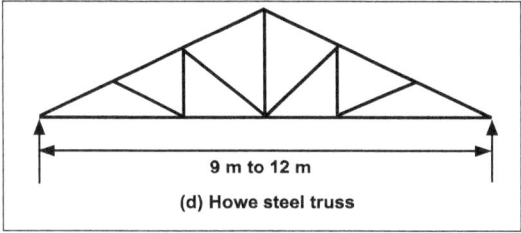

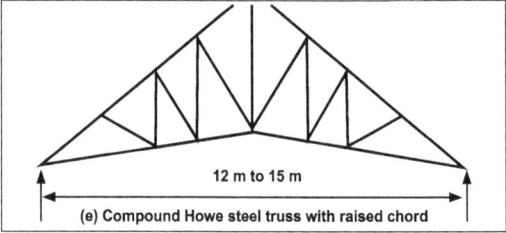

Various types of steel trusses for spans up to 15 m.

Small trusses (span up to 15 m) are rested on bed plates at the ends. Bed plates may be of stone or concrete. Ends are bolted down with rag bolts which holds the truss down.

Small trusses generally consist of angles connected with gusset plates. To seat the foot of the truss on the bed plate short angles are fitted. For spans up to 7 m, 15 mm diameter rivets are used (Figure a).

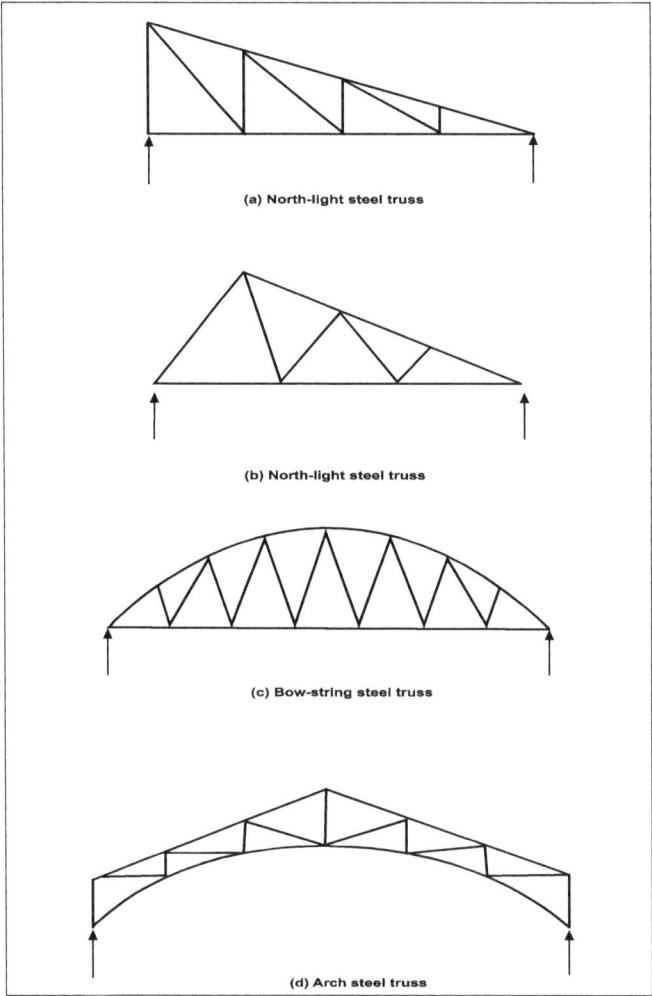

(a) North-light steel truss

(b) North-light steel truss

(c) Bow-string steel truss

(d) Arch steel truss

Various types of steel trusses for span more than 15 m.

2. Trusses for Large Spans

In large-span trusses, the members may consist of more than one section. Two angles or channels or flats may be connected in the gusset with other members at the joints.

As the truss is hung, handling may be difficult. Trusses are riveted into two portions at the fabrication centre and gusset plates at the connected ends are riveted to them.

Two halves are erected at the site and riveted. Bigger bearing plates are used for these trusses.

3. Fabrication of Steel Work

Preparation of steel work for erection is called as fabrication. It includes all work necessary to layout, cut, drill and rivet or welds the steel sections.

The fabrication plan has to be properly planned such that the work at the site of erection should be reduced as much as possible. Most of the work is carried out in the fabrication shop itself.

All materials procured from the mill or the markets are straightened if necessary. Cutting is affected by shearing, cropping or sawing.

For mild steel, gas cutting by mechanically controlled arcs is also used. For high tensile steel, gas cutting is permitted under special care. Plates and angles are cut by shearing.

Beams and channels are usually cut to the desired lengths in factories. Templates according to the shape of the final job are prepared. Templates may be made of wooden strips showing location of all holes and cuts.

Card -board templates may be used for gusset plates. All the materials are laid out. The centers of the holes are identified with a punch. Holes are drilled, punched or bored.

Drilling is generally preferred as it gives the exact hole without spoiling the surroundings. For thick sections drilling is preferred.

When the components of a member are ready they are held in position temporarily by shop bolts. Shop bolts are longer in size than the normal ones. There should be atleast tow bolts put in one part of a member.

All assembled parts should be in close contact. All bearing stiffeners should bear tightly at top and bottom. In general, no drifting should be permitted.

Assembled parts are then riveted adopting hydraulic or pneumatic pressure. Rivets of diameter less than 10 mm are driven cold.

Bigger rivets are heated by a steady flame produced by burning oil or using electric heaters. It is the practice that the rivet fills the hole fully and forms a head of standard size.

After finishing the required structure formation, it is cleaned thoroughly and one coat of red lead paint is given. All machined faces should be coated with a mixture of white lead and tallow.

All inaccessible portions of the structure should be painted with two coats of red lead paint. Completed work should be temporarily shop erected such that accuracy of fitness may be checked before dispatch.

4. Erection of Steel Work

Structural steel work is erected with the use of derricks, slings, guys, cranes, etc. The erection is done based on the drawings and keeping the verticality of the columns and fixing other members and parts of the steel structure.

During erection, the steel work should be temporarily braced till the final stage is reached to allow the structure to take the required load.

All the riveting or welding works, if any, is done after proper position is attained. Final alignment and verticality are checked. The steel work is finally painted.

2.21 Frames

Frame structures are the structures having the combination of beam, column and slab to resist the lateral and gravity loads.

These structures are usually used to overcome the large moments developing due to the applied loading.

Types of Frame Structures

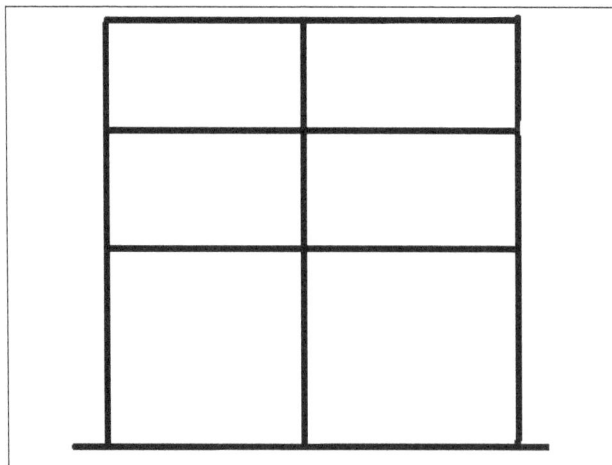

Frames

Frames structures can be differentiated into:

1. Rigid Frame Structure

Which are further subdivided into:

- Pin ended.
- Fixed ended.

2. Braced Frame Structure

Which is further subdivided into:

- Gabled frames.
- Portal frames.

Rigid Structural Frame

The word rigid means ability to resist the deformation. It is defined as the structures in which beams & columns are made monolithically and act collectively to resist the moments which are generating due to applied load.

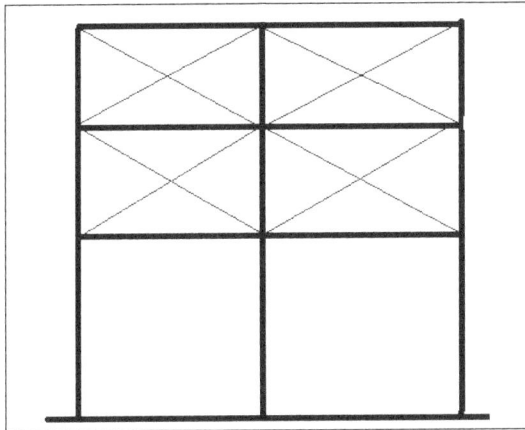

Rigid Structural Frame.

It provides more stability. This type of frame structures resists the shear, moment and torsion more effectively than any other type of frame structures.

Braced Structural Frames

Here, bracing are usually provided between beams and columns to increase their resistance against the lateral forces and sideways forces due to applied load.

Bracing is usually done by placing the diagonal members between the beams and columns.

It provides more efficient resistance against the earthquake and wind forces. It is more effective than rigid frame system.

Pin Ended Rigid Structural Frames

It has pins as their support conditions. Considered to be non-rigid if its support conditions are removed.

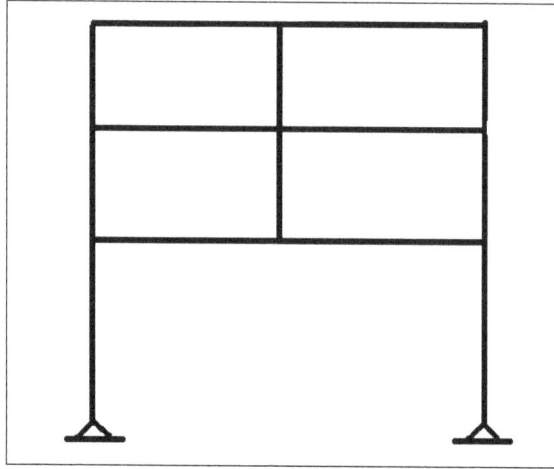

Pin Ended Rigid Structural Frames.

Fix Ended Rigid Frame Structure

In this type of rigid frame systems, end conditions are usually fixed.

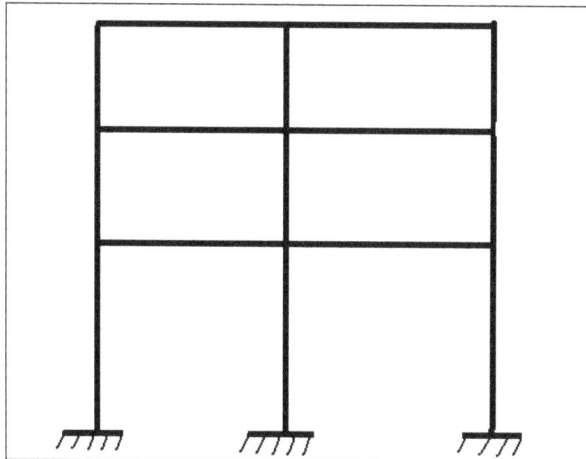

Fix Ended Rigid Frame Structure.

Gabled Structural Frame

It usually has the peak at their top. Used where there are possibilities of heavy rain and snow.

Portal Structural Frame

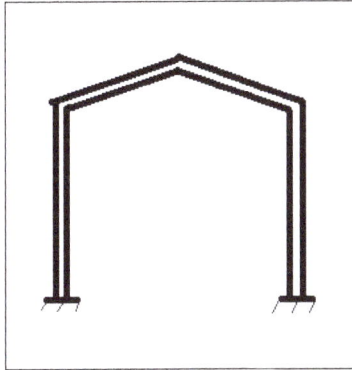

Portal Structural Frame.

It usually looks like a door and used in construction of industrial and commercial buildings.

Load Path in Frame Structure

It is a path through which the load of a frame structure is transmitted to the foundations.

In frame structures, usually the load path is the load that first transfers from slab to beams then to from beam to columns, then from columns it transfers to the foundation.

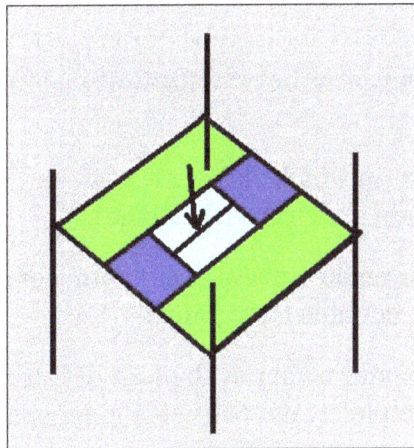

Load path in Frame Structure.

Advantages of frame structures:

- Easy to construct and it is very easy to teach the labor at the construction site.

- Frame structures can be constructed rapidly.

- Frame structures have economical design.

Disadvantages of Frames:

- Span lengths are usually restricted to 40 ft when normal reinforced concrete.

2.22 Braced Domes

Vast majority of domes of a larger span are of this type. It has special importance in engineering practice.

Composed either of members lying on a surface of revolution or of straight members with their connecting points lying on such a surface.

In three-dimensional structure, any applied loading is distributed between many members which may be at a considerable distance from the point of application of the load.

This leads to an even stress distribution in the structure, as the high stresses in the directly loaded members are decreased and the stresses in the more distant members are increased.

Braced domes can be subdivided into:

- Frame or skeleton type (single-layer dome).

- Truss type (or double-layer dome), extremely rigid and suitable for very large spans.

- Stressed skin type (in which the covering forms an integral part of the structural system).

- Formed surface type (in which sheets arc bent and interconnected along their edges to form the main skeleton of the dome).

Modern braced domes are often prefabricated consisting of a small number of different members interconnected by suitable node joints.

They possess great stiffness and owing to the lack of internal columns or horizontal bracing members have a completely unobstructed inner space.

2.23 Laying Brick

It is both an art and a science. Foundations of good bricklaying are the same for any project. Some of the essential skills for this task are also relevant to paving and tiling projects.

Laying of bricks.

Basic Bricklaying Equipment

Gather our tools and materials before we start laying bricks. Once we have mixed the mortar any that isn't used while still fresh will go to waste.

These are the essentials:

- Brick trowel.
- Spirit level.
- Jointing bar.
- Tape measure.
- String.
- Soft brush.
- Mortar mixture.
- Mortar board.
- Mixing bucket, wheelbarrow or cement mixer.

Mixing Mortar

Mortar is made from cement, sand, water and a mortar additive. We can use a cement mixer or mix it by hand using a shovel and a large container such as a wheelbarrow.

Different quantities of sand and concrete are used for varying purposes and by different bricklayers. Most common proportions typically require four parts sand to one part cement or three parts sand to one part cement.

Laying Bricks

Scoop up some mortar with a trowel and start at one corner of our construction. Lay the mortar down an inch thick and a few inches wide, then set a brick down on top of the mortar and tap it into position with the handle of the trowel until it lines up correctly.

Use our spirit level to check that it's aligned horizontally, as well. Next, use our trowel to scrape away excess mortar from the joint and place our next brick in the same manner. Frogged bricks must always be laid frog-up so that the frogs are filled with mortar.

A mortar joint should be roughly 3/8 to 3/4 inches thick and all our joints should have an equal thickness. Build up height at the ends or corners of our construction, then pin string taut between these end bricks to guide us in laying bricks level in the middle sections.

2.24 Weather and Water Proof

Most of the leaks occur at window and door openings or at intersections between building components. In some cases, caulks and sealants forestalled leakage at these poorly designed joints for the first few years.

Eventually most caulk joints fail, allowing water to enter. All residential cladding systems are more or less porous to water, particularly during wind-driven rain when high air pressures on the windward side of a building force water to flow towards lower-pressure areas behind the siding.

Water exploits.

Under pressure, the water exploits butt joints, lap joints, nail holes and other openings to flow inside (Figure). Even without wind, some water will migrate through tiny gaps to the back of siding through capillary action, the way water is siphoned up a stalk of celery. This is true of brick, wood and stucco, as well as the newest composite materials.

Water penetration of an older home siding.

In older construction, water that penetrated the outer cladding had ample opportunity to dry both to the interior and to the exterior as wind washed through the wall cavities, which were kept warm by heat leaking from the building's interior.

In modern construction, however, with high levels of insulation, continuous air and vapor barriers and low-perm sheathing panels, when water gets in, it is much slower to dry and more likely to cause damage.

This structure relied on diagonal bracing for stiffness rather than an exterior sheathing board. Later an insulation improvement included blowing cellulose into the building walls - which was fine.

But now, water that used to leak into the wall cavity during windy rainy weather soaked the wall interior and was more of a problem.

Luckily cellulose insulation, probably because of the chemistry of its fire-retardant treatment, is rather mold-resistant. But that doesn't necessarily prevent an attack by termites or carpenter ants.

Leaks behind vinyl siding form ice freezing climates. Ice can also show where water leaks or moisture problems are occurring.

Leaks behind vinyl siding form ice freezing.

While the exterior finish should be detailed to repel and shed water, a backup system is needed for the times when the primary system fails.

The backup system needs to catch any water that penetrates the cladding and to drain it safely to daylight at the bottom of the wall.

The source of this leakage needs to be found and cured to avoid costly problems such as structural rot, insect damage and even a wall cavity mold contamination issue.

The backup layer in an exterior wall, called a water-resistive barrier by the International Residential Code (IRC), typically consists of properly lapped building paper or plastic house wrap integrated with all flashing to safely drain water away.

It is also called the drainage layer or drainage plane. In this approach, the outer cladding functions as a decorative "rain screen," slowing down wind and water, but it is not expected to be 100% waterproof.

2.25 Roof Finishes

Roof is the upper most part of a building which is constructed on structural members and provided with a cover to protect the building and the occupants from atmospheric effects like sun, wind, rain, snow, etc.

Requirements of a Roof:

- It should give a protective covering against the adverse effects of atmosphere.

- It should have adequate slope to drain the rain and snow.

- It should be durable and strong enough to take the external loads.

- It should provide sufficient insulation against sound.

- It should meet the different climatic and covering materials available.

- It should also provide good architectural appearance.

Types of Roofs

Roofs may be grouped under two major categories, viz., sloping or pitched roofs and flat roofs.

1. Sloping or Pitched Roofs

Sloping or pitched roofs are those which have the surfaces with considerable slope for covering the building structure. These roofs are constructed out of wood, steel or combination of both.

These roofs are generally lighter in weight than flat roofs. These roofs are mostly used in regions of heavy rainfall or snow fall.

Different types of roof coverings are in use depending on the type of pitched roof and the availability of material.

2. Flat Roofs or Terrace Roofs

A roof which is nearly flat is known as a flat roof. It is the convention if the slope is less than 10°, it is considered as a flat roof.

As a matter of fact no roof is laid flat. But flat roofs are normally constructed with a slope of 1 in 2 to 1 in 6.

Slope is provided in one or more directions such that the rain water is drained off rapidly. Flat roofs have more advantages compared to disadvantages.

Special care has to be taken with reference to drainage, weather and water proofing. In order to meet, these needs necessary roof finishes have to be adopted.

Drainage of Flat Roofs: It is one of the basic requirements of flat roofs. The rain water should be drained off quickly so as to avoid leakage of roof. In order to achieve this satisfactorily water-tight roof surfaces, adequate slope and drain-outlets have to be provided.

The slopes in flat roofs vary from 1 in 20 to 1 in 40. The rain-water outlets should be evenly spaced round the building. Some possible arrangements of drain-outlets are shown in figure.

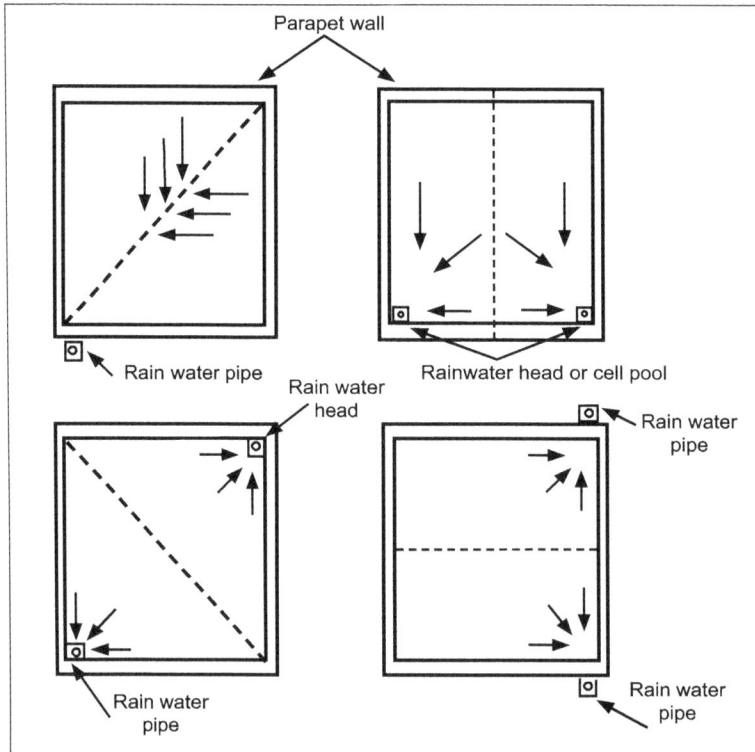

Different arrangements of drain-outlets.

Water-Proofing of Flat Roofs

Most of the flat roofs now-a-days are of reinforced cement concrete. The flat roof should be made of water-proof by a surface covering. The following methods are employed:

- Cement mortar finishing.
- Bedding concrete and flooring.
- Mastic asphalt.
- Water-proofing compounds.

Cement mortar finishing: For small size buildings of normal use, the finishing of roof surface is done at the time of laying cement concrete. The flat roof surface is finished with cement mortar of 1: 4.

Bedding concrete and flooring: The concrete surface is kept rough over which brickbats lime concrete of 1: 2: 4 or brickbats cement concrete of 1: 8: 14 is laid for a thickness of about 10 cm.

The bedding concrete is finished with tiles, stone slabs or terrazzo, etc. A convex joint is provided at the junction of parapet wall and the roof.

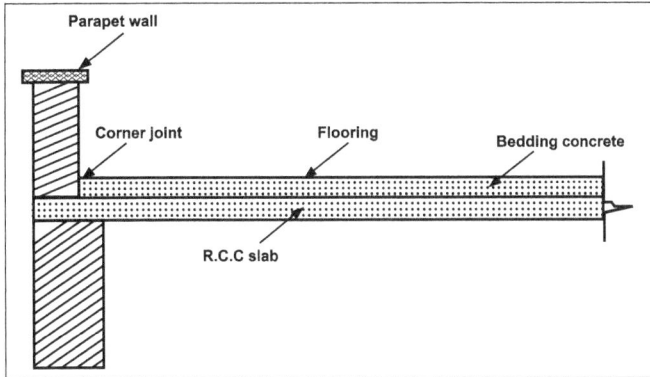

Water-proofing by bedding concrete in flat roof.

Mastic asphalt finish: A layer of hot mastic asphalt is laid on the roof surface and a jute cloth is spread over this surface.

One more layer of mastic asphalt is laid over the jute cloth such that the jute cloth is sandwiched between two asphalt layers. Then sand is sprinkled over the entire surface of the roof.

A lead sheet is inserted at the junction of the parapet and the roof so as to have a better grip and easy draining.

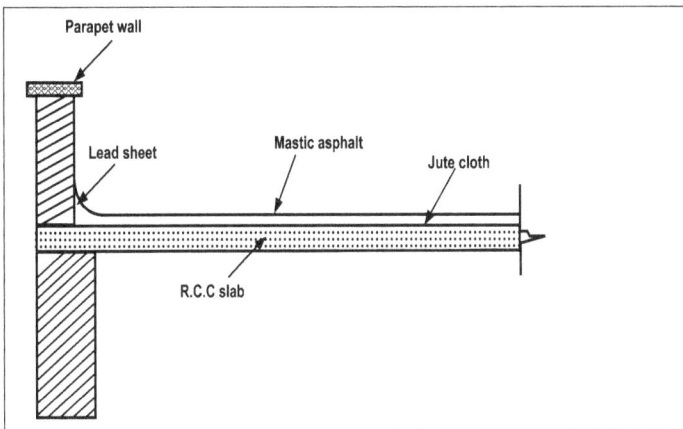

Water proofing by mastic asphalt in flat roof.

Water-proofing compounds: Commercially available water-proofing compounds are added to cement during construction. These compounds prevent seepage, leakage and dampness caused by capillary absorption of moisture in concrete. Only 2 % of the compound is added.

Weather-proof course for flat roofs: Flat roofs are provided with a weather-proof course to sustain the effects of atmospheric agencies. The construction of weather-proof course consists of providing one course of surki-concrete and two courses of flat tiles.

The surki-concrete (broken brick aggregates and lime with no sand) in the mix ratio of 1: 2: 5 is laid to a compacted thickness of 75 mm.

The concrete layer is provided with a minimum slope of 1 in 50 directed towards the rain water drains. After curing for six days, two courses of flat tiles or one course of pressed tile are laid in a cement mortar 1: 3 with crude oil.

2.26 Acoustic and Fire Protection

Sound is produced when part of the atmosphere is compressed suddenly. It is transmitted in the form of waves. The waves are a series of compression and rarefaction created in the air medium.

The average sound travels at a speed of 340 m per second at ordinary temperature. It depends on the medium through which it travels.

Effects of Audible Sound

Sound can travel through some medium like air. It cannot travel in vacuum. Thus, for the sound to be audible to the ears, the sound source and ear must be connected by an elastic medium like air. Following are the characteristics of audible sound:

Frequency of Sound: Frequency of pitch of sound is defined as the number of cycles or vibrations per second.

The highest audible sound (e.g., whistle) has a frequency of 20000 cps (cycles or vibrations per second).

The lowest audible sound has a frequency of 20 cps (e.g., whispering). The frequency is a measure of the quality of sound.

Intensity of sound: The intensity of sound is defined as the flow of sound energy per second through unit area. The intensity of sound is the strength of the sensation received by the human ear.

Intensity of sound is a purely physical quantity. But loudness of sound depends on the characteristics of ear.

Measurement of sound: The range of intensity of sound is very large. The loudest sound is about 10_{13} times the sound which is just audible by the human ear. There is a wide range of sound levels, it is realized that a scale has to be adopted as a guidance.

The intensity of sound is measured on a logarithmic scale due to wide range of variation of the intensity of sound. Bel is the measure of intensity of sound named after Graham Bell, the inventor of telephone.

As the unit of bel is comparatively large, hence a shorter practical unit decibel (db) equal to 1/10 of a bel is used.

The range of audible sound too painful noise varies from 1 to 10_{13} which is covered on logarithmic scale between 1 to 130 db units.

One db unit is approximately the smallest change of sound intensity which the human ear can hear.

Principles of Acoustics

The behaviour of sound plays an effective role in the acoustical design of different type of buildings and in the sound insulation process.

A sound originating from a source, such as music or operation of machine, is transmitted through the medium in all directions. The transmitted sound strikes on some surface, like wall, ceiling, floor or any other barrier.

Depending on the type of surface part of it is reflected back and a part being absorbed by the surface. If the sound is not absorbed by the material, it will be transmitted in part to another side of barrier.

If the ultimately reflected sound is not properly controlled the reflection may result in acoustical defects, viz., echoes and reverberation. This reflected sound is important in the acoustical design of buildings.

The part of sound absorbed by the surface is represented by an absorption coefficient. This coefficient is the ratio of the energy absorbed by the area of the surface to the energy striking the area. This coefficient is a function of the frequency of sound.

The reduction of intensity of sound of a transmitted sound through a barrier is called as transmission loss. This transmission loss is a measure of the effectiveness of a surface as an insulating material.

Thus, transmitted and absorbed sound have important bearing on the acoustical condition of a building. However, both transmitted and absorbed sounds are inter related and influence the acoustic and sound insulation.

Highly porous materials have the quality to dissipate considerable energy and the absorption will be relatively high.

Acoustical Defects

The acoustical design of an enclosed space is basically depend on the behaviour of the reflected sound. Due to the reflection of sound two main defects are developed, viz., echoes and reverberation.

Echoes: Echo is said to be produced when a reflected sound wave reaches the ear, just when the original sound from the same source has already been reached.

The sensation of sound persists for one-tenth of a second after the source has ceased. Thus, an echo forms when the time lag between the two sounds is about 1/17th of the second.

Further considering the velocity of sound in the atmosphere air as 34.3 m/sec., it is shown that when the distance of the reflecting surface is between 8 m and 17 m, echoes are formed.

The defect of echoes also occurs when the shape of the reflected surface is curved with smooth character. Echoes are unpleasant to hear and cause disturbance in hearing.

The methods of reducing this defect are choosing proper shape of the surface, a rough and porous material to disperse the energy of echoes.

Reverberation: When the surfaces of an indoor place are hard and smooth, very small energy is lost at each impact of sound and many reflections take place before the sound dies down.

This repeated reflection of sound is called prolongation or reverberation. Thus, if sound exists too long, then successive words of a speech will overlap and confuse.

The remedy for this defect is to select a correct time of reverberation known as optimum time of reverberation. This is achieved by suitably selecting a proper absorbent or acoustical material for different reflecting surfaces.

Acoustic levels of a room and reverberation time are shown in table:

S. No	Acoustics level	Reverberation time in seconds
1.	Excellent	0.50 to 1.50
2.	Good	1.50 to 2.00
3.	Fairly good	2.00 to 3.00
4.	Bad	3.00 to 5.00
5.	Very bad	Above 100

Acoustical Materials

Common building materials are absorbents of different levels. Such materials are called as absorbent materials.

Some of the acoustical materials are as follows:

- Acoustic Plaster.

- Acoustic Tiles.

- Porous Boards.

- Perforated Boards.

- Quilt and Mats.

Fire Protection

When some materials get ignited, the material catches fire and spreads. If there are openings in walls and floors, the fire spreads to more areas.

If there are no openings, the temperature of the structure is increased by fire. In buildings, staircases and lift shafts act as flues for fire and increase the possibility of spreading of fire.

There are natural and man-made causes for fire to occur. They may be caused due to faulty workmanship in electrical wiring, leakages in heating and cooking equipment, flammable liquids, careless throwing of cigarette bits and matches, lightening, spontaneous combustion, etc.

The fire spreads over different materials and produces different gases of which some are poisonous.

The gases produced are carbon-monoxide, carbon-dioxide, hydrogen sulphide, nitrogen dioxide, etc. Thus to protect the goods and activities within a building or structure and of adjacent buildings fire-protection has to be resorted to.

Fire resistance of a material is the time during which a structure fulfills its function with reference to safety when a fire presents with a particular intensity.

Fire-resisting Properties of Building Materials

With reference to fire, materials may be classified as combustible materials and non-combustible materials.

Combustible materials are the materials which combine exo-thermally with oxygen and give rise to flame at a particular range of temperature.

Examples of such materials are wood, wooden products, animal products and man-made products like fibre board, straw-board, etc.

Non-combustible materials are those which when decomposed by heat will do so endo-thermically. These materials do not catch fire by one decomposed at a particular range of temperature.

Examples of such materials are metal, stone, glass, concrete, clay products, gypsum products, asbestos products, etc.

The building materials have varying fire-resistant properties which are discussed below:

1. Bricks

Bricks in general have good fire-resistant property. Particularly first class bricks are fire-proof and can withstand heat for a considerable length of time.

As bricks are made out of clay, which is a poor conductor, can withstand heat as high as 1300°C. Special types of bricks called fire-bricks are best for use in fire-resistance constructions.

2. Terra-cotta

Terra-cotta is also a clay product which has better fire-resisting properties than bricks. As the cost is high it is used only in restricted places.

3. Stone

Although stone may resist high temperature but deteriorates due to sudden cooling. Thus, stone should be used only for a limited use in buildings with reference to fire-resistance. Granite although very strong crumbles or cracks when subjected to heat.

4. Concrete

As concrete is a bad conductor of heat, it has high fire-resistance capability. The extent of fire resistance depends on the aggregate, density and position of reinforcement in RCC. Use of foamed slag, blast furnace slag, crushed brick, cinder, crushed limestone, etc., form the best aggregate for fire-resisting concrete.

5. Mortar

Mortar is a cheap and best incombustible material. Cement mortar is better fire-resistant than lime mortar as lime plaster is susceptible by calcination.

To increase the fire-resistant property, the thickness may be increased.

Cement mortar with surki or pozzolana shows very high fire-resistance capability.

Fire-Load

The amount of heat liberated in combustion of any content or part of a building of a floor area is referred to as fire-load. It is represented in kilo joules per square metre (kj/m²).

The fire-load is the ratio of the weight of all combustible materials to the floor area under consideration.

For example, let a floor area of 120 m² contain 18×10^3 N of combustible material having caloric value of 1.5×10^3 j/N, then the:

$$\text{Fire} - \text{load} \frac{18 \times 10^3 \times 1.5 \times 10^3}{120} = 225 \times 10^3 \, \text{j/m}^2$$

The fire-load is used as a measure of grading of occupancies by Bureau of Indian Standards (BIS 1641-1960).

Accordingly the classifications are given as:

- Low fire-load.

- Moderate fire-load.

- High fire-load.

Table: Grading of occupancies by fire-load (BIS 1641-1960).

SI. No	Class of fire	Limit in kj/m²	Occupancies load
1.	Low	Not exceeding 1.15×10^6	Domestic buildings, hotels, boarding houses, restaurants, schools, hospitals, temples, mosques, offices, factories where NH materials are used, etc.
2.	Moderate	1.15×10^6 to 2.30×10^6	Retail shops, emporium, markets factories, workshops, etc.
3.	High	$2.23 \text{ xz } 10^6$ to 4.60×10^6	Godowns and similar structures used for bulk sto rage of NH materials and goods.

General Safety Requirements against Fire

All buildings should satisfy certain safety requirements against fire, smoke and fumes.

1. Maximum Height

The height of a building is restricted depending on the number of storeys, the number of occupancy and the type of construction.

Further all the above factors in turn depend on the width of the road in front of the building, floor-area ratio and the local firefighting facility available.

2. Open Space

In general, every room for use by human beings should have an interior or exterior open space or on an open verandas.

The open spaces inside or outside should be able to provide sufficient lighting and ventilation. Further, the open space in adjoining a road should be well inside giving scope for widening of the road.

3. Mixed Occupancy

When a building is used for more than one type of occupancy, e.g., residential, godown, shops, etc. then it should conform to the requirements for the most hazardous of the occupancies. Such mixed occupancy should be avoided as there is more risk for life of occupants.

If mixed occupancy is separated by walls of 4-hour fire resistance, then the occupancy can be treated individually and safety measures can be taken.

4. Openings in Separating Walls and Floors

The openings in separating walls and floors should be designed in such a way that necessary protection is guaranteed to all such factors which may spread fire.

For Types 1 to 3 construction a door way or opening in a separating wall may be limited to about 6 m² (i.e., height 2.75 m and width 2.1 m). Such wall openings should be provided with fire-resisting doors or steel rolling shutters.

All openings in the floors shall be protected by vertical enclosures. In Type 4 construction, openings in the separating walls or floors should be fitted with 2 hour fire-resisting assemblies.

5. Enclosure on Openings

Wherever openings are permitted, they should not exceed three-fourths the area of the wall in the case of external wall and should be protected with fire-resisting assembles or enclosures. Such assembles and enclosures shall also be capable of preventing the spread of fumes or smokes.

6. Power Installations

Electrical power installations and gas connections for kitchen, if any, should be done as per norms and requirements from the point of view of fire safety.

7. Materials of Construction

The structural elements of the building such as floors, partitions, roofs, walls, etc. should be invariably constructed with fire-resisting materials.

In general, non-combustible materials like stones, bricks, concrete, metal, glass, clay products, etc. should be used in construction.

Combustible materials such as wood and wood-products, fibre boards, straw boards, etc. should be avoided or used only for the most essential places.

Emergency Fire-Safety Measures

Apart from the steps taken in construction of buildings, the following general measures of fire-safety have to be adopted. It includes the following:

- Alarm Systems.

- Fire-extinguishing Arrangements.

Various types of extinguishing arrangements are provided to extinguish the fire depending on the importance of the building. They are as follows:

- Portable Fire-extinguishers.

- Fire Hydrants.

- Automatic Sprinkler System.

- Escape Routes.

Sub-Structure Construction

3.1   Techniques of Box Jacking

Box jacking is jacking a large precast reinforced concrete box horizontally through the ground, usually beneath a road or railroad that must not be interrupted. The major advantage of the process is its simplicity. Only the exact prism of earth that will be filled by the jacked box is excavated. No intermediate ground supports are needed.

The structure is built away from the roadway without the constraints of shoring and traffic controls. When the structure is ready, a shield is fitted to the front, hydraulic jacks are installed behind and the box is pushed into final position while simultaneously the earth is excavated from within.

The actual jacking generally takes only a few days to a week. During that time, traffic is proceeding overhead normally, unaware of the construction below. The non-disruptive nature of the process together with its inherent safety, simplicity and economy make box jacking a useful tool for the practicing civil engineer.

Applications

Some examples of potential box jacking projects include storm drains, bike or pedestrian trails, livestock or wildlife under crossings, conveyors, pipe ways and other industrial uses, small bridges and roadways up to 4 lanes wide. Basically, applications of box jacking depend only on the creativity of the civil engineer designing the project.

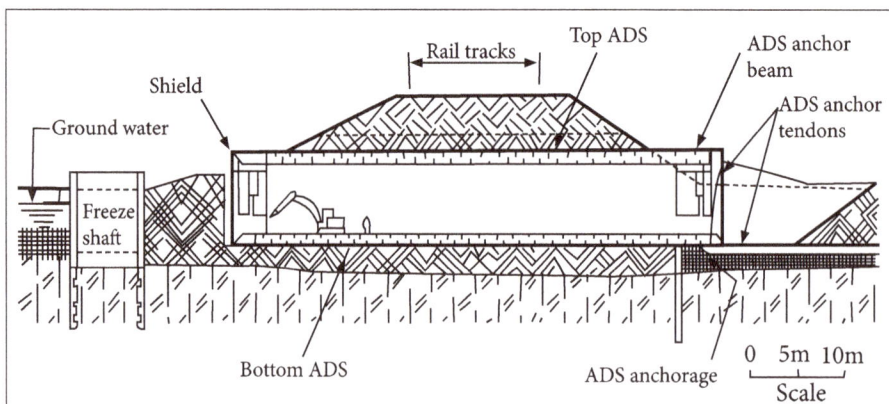

Longitudinal section.

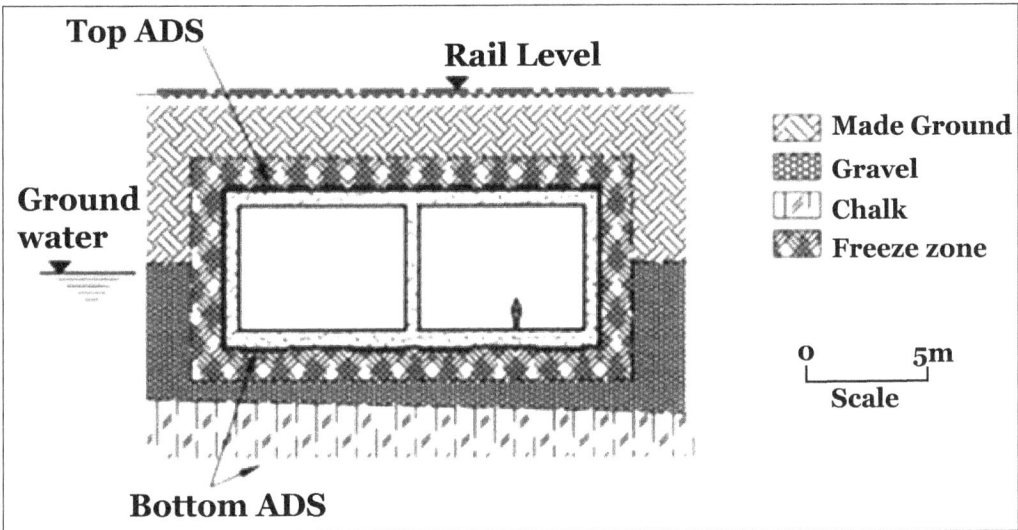

Top ADS

Rail Level

Ground water

- ▨ **Made Ground**
- ▦ **Gravel**
- ▨ **Chalk**
- ▨ **Freeze zone**

0   5m

Scale

Bottom ADS

Cross section (Box jacking)

3.2   Pipe Jacking

With a pipe jacking, utility lines and pipes are driven underground in segments from a starting shaft, using hydraulic jacks. After the starting and target shafts designed to absorb the jacking forces have been constructed, the face is cut with a cutting shoe or under the protection of a shield tunneling machine.

With each successive pipe, the cutting shield is advanced, the hydraulic jacks bearing against the reinforced wall of the shaft. In this way, each individual pipe is lowered into the shaft, joined to the previous one and then jacked forward.

The other cut produced by the cutting shoe is grouted with cement through openings in the pipes. Intermediate jacking stations are used to reduce the jacking pressures and facilitate curve drives.

With longer tunnels, there are intermediate shafts which may be crossed by the pipe section. The pipes are usually made of high-strength concrete to withstand the high jacking forces.

They form driving elements and permanent supports at the same time. Their wall thickness and weight are determined by the maximum compression forces occurring. They can be up to 40 cm thick and weigh more than 50 tones and they are done on or off site.

The term pipe jacking can be used to describe a specific installation technique as well as a process applicable to other trenchless technology method. While referring to a process it implies a tunneling operation with use of thrust boring and pushing pipe with hydraulic jacking force.

This concept of jacking system is adopted by many trenchless technologies, including auger boring and micro tunneling. However, the term pipe jacking is regarded rather as an installation technique.

Pipe jacking is a trenchless technology method for installing a prefabricated pipe through the ground from a drive shaft to a reception shaft. In the pipe jacking operation, jacks located in the drive shaft propel the pipe.

The jacking force is transmitted through the pipe-to- pipe interaction, to the excavating face. When the excavation is accomplished, the spoil is transported through the jacking pipe in the drive shaft by manual or mechanical means.

Both excavation and spoil removal processes require workers to be inside the pipe during the jacking operation.

This is essentially what separates pipe jacking from micro tunneling. Although it is theoretically possible for a person to enter a 36-inch diameter pipe the minimum recommended diameter for pipe installed by pipe jacking is 42 inches.

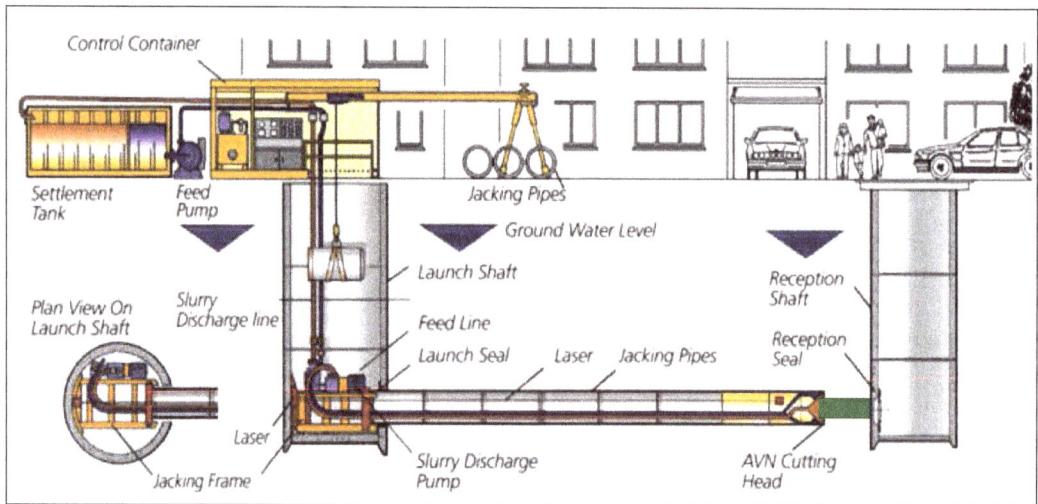

Pipe jacking.

3.3   Under Water Construction of Diaphragm Walls and Basement

Diaphragm Wall is generally reinforced concrete wall constructed in the ground using under slurry technique which was developed in Europe. The technique involves excavating a narrow trench that is kept full of an engineered fluid of slurry. Walls of thickness between 300 and 1200mm can be formed in this way up to depths of 45 meters.

Diaphragm Wall Application

- Commonly used in congested areas.

- Practically suited for deep basements.

- Used in conjunction with "Top Down" construction technique.

Positive Facades of Diaphragm Wall

- Can be installed to considerable depth.

- Formation of walls with substantial thickness.

- Flexible system in plan layout.

- Easily incorporated into permanent works.

- Designable to carry vertical loads.

- Construction time of Basement can be lowered considerably.

- Economic and Positive solution for large deep basement in saturated and unstable soil profiles.

- Can be used for seepage control in Dams.

- Noise levels limited to engine noise only.

- No vibration during installation.

Negative Facades of Diaphragm Wall

- Not economical for small, shallow Basements.

Category of Diaphragm wall

- In Situ Cement Bentonite Vertical Wall.

- In Situ RCC Vertical Wall.

- Precast RCC Vertical Wall.

In Situ Cement Bentonite Vertical Wall

- Provides water tight barrier.

- Used to prevent seepage/water loss from Natural reservoir and Dams.

In Situ RCC Vertical Wall

- Used for Retention systems and Permanent foundation walls.

- Used for deep groundwater barriers.

Trenching Equipments used are:

- Hydraulic Grab.

- Kelly-mounted or Cable-hung cam buckets.

- Mechanical Grab.

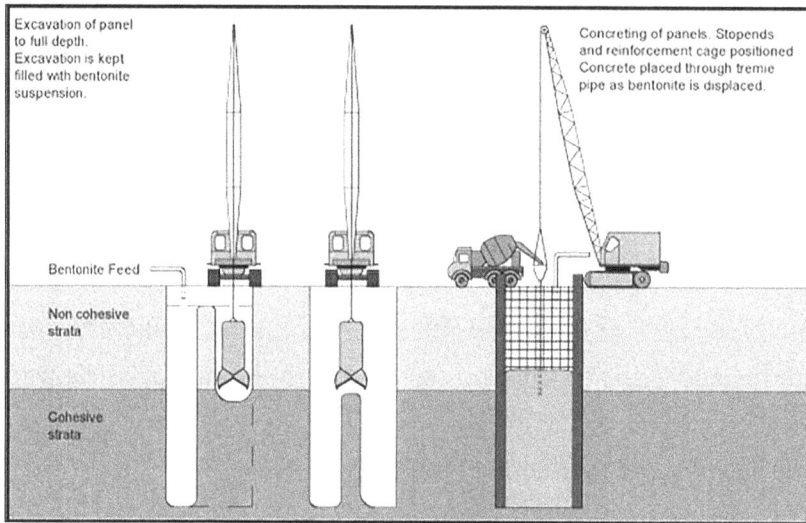

Diaphragm walls.

3.4   Tunneling Techniques

Methods of Tunneling

The choice of a particular method of tunneling depends on the type of ground.

The types are as follows:

- Firm ground.

- Soft ground.

- Running ground.

- Rock.

Tunneling in Firm Ground

In firm ground, sufficient period is available for installing conventional support. Further, the method to be adopted depends on the shape, size and available equipment.

1. Full Face Method

This method is suitable in comparatively firm soil where the excavated portion can hold itself for sufficient time to permit mucking and supporting operations to be completed. Here, the proposed cross-section of the tunnel is excavated in comfortable sections.

The excavation to be done is divided into three or more sections. First the top section I is cut and removed. This is followed by cutting sections II and III in turn. This method is recommended for tunnels of small size.

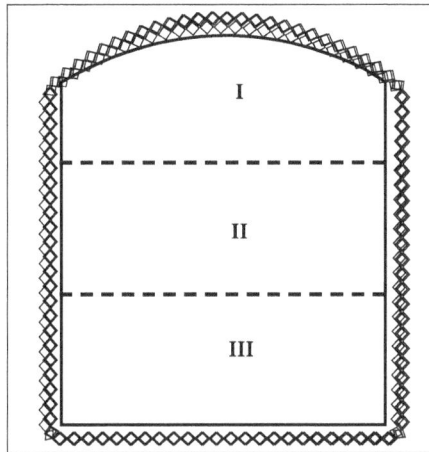

Full face method.

2. Top Heading and Benching Method

This method is adopted when the excavated portion cannot hold itself till mucking and supporting operations are carried out. So, the heading is excavated first and supported to the full length or part length of the tunnel before benching is started.

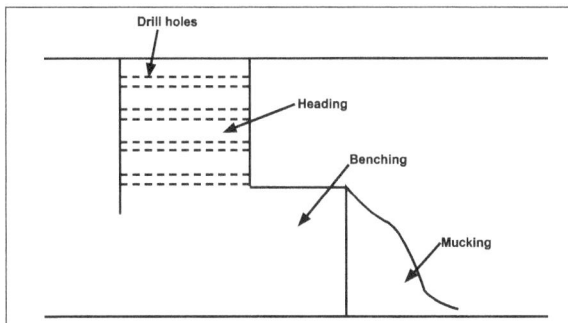

Heading and benching method.

The heading is always ahead of benching by a convenient length. This may be formed by excavating full width of the tunnel above the springing line. The principle of the method is shown in figure.

3. Drift Method

In the case of large size tunnels, a pilot tunnel or drift is made in the side or at the center of the tunnel. Drill holes are driven from the drift towards the periphery and drift widened. The drift provides suitable arrangement for supporting the excavation. The drift location depends on the type of tunnel; accordingly the methods are classified as wall plate drift, side drift and multiple drift.

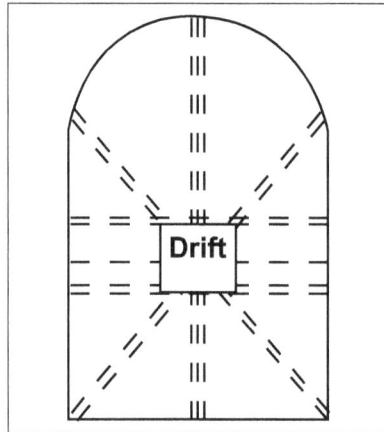

Drift method.

Tunneling in Soft Ground

Instantaneous support is needed in case of soft soil before drilling and blasting. In such cases, the traditional method adopted is fore poling method. This method consists of driving boards ahead to support the ground ahead of rib which are known as spiles. The forepoles act as cantilevers beyond blasting and carry the weight of the ground. They carry till the forward ends are supported by the steel rib.

The spiles are installed as far around the periphery as necessary. After removing the breast boards and the new rib is erected in position and then the soil is excavated. Afterwards breast boards are fixed and the operation is repeated.

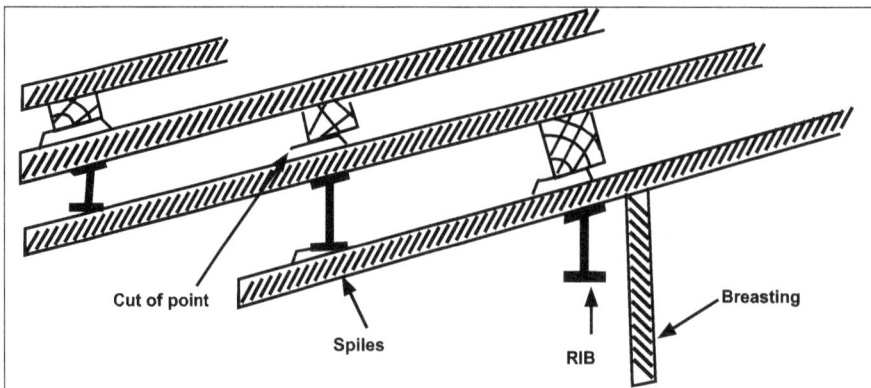

Fore poling method of tunneling.

Tunneling in Running Ground

In this type of ground, special treatment has to be resorted to before starting the excavation. Following methods are adopted:

1. Tunneling with Liner Plates

On medium stiff ground, this method is employed for driving steel lined small section drifts or headings. The first liner plate is kept at the crown in a pre-excavated cavity. Two adjacent liner plates are bolted to it one on either side after widening the hole.

These plates are supported by trench jacks or by props carefully tightened. Then, the arch section is widened gradually down to the springing line. This arrangement in combination with stiffener rings is suitable for use in very large tunnels.

2. Needle Beam Method

In this method, full section of the tunnel is broken out. At the time of excavation, plates are placed one by one. These plates are supported by radially set trench jacks from a centrally placed longitudinal girder called needle beam.

The needle beam is kept at the bottom of the top heading. After placing the beam, the trench jacks are removed. Concreting is done at top and bottom.

Tunneling in Rock

Tunnels in rock are driven by repeating the following sequence of operations:

- Drilling hole on the rock face.
- Loading the holes with explosives.
- Blasting.
- Removing the debris.
- Disposing off the broken rock.

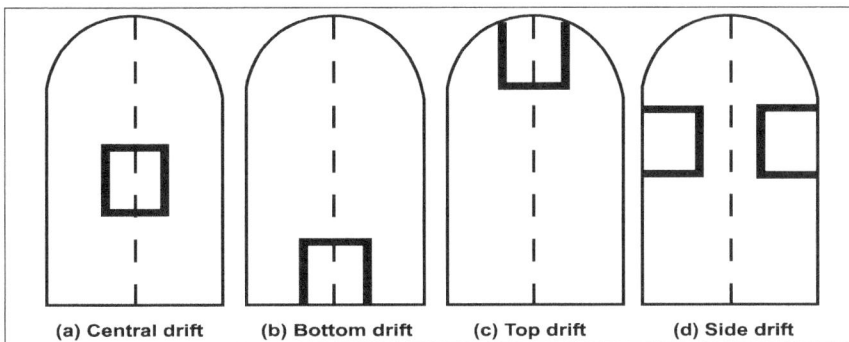

(a) Central drift   (b) Bottom drift   (c) Top drift   (d) Side drift

Different positions of drift.

In each sequence, full cross-section of the tunnel may be excavated. Instead one or more drifts may be excavated in advance. The following methods are adopted:

Tunnel Lining

Tunnel lining is intended to withstand the following loadings:

- Weight and pressure of surrounding ground.
- Water pressure.
- Super imposed loading.
- Service loading such as load related during use, viz., railway, highway, etc.
- Temporary load during construction such as due to compressed air pressure.
- Weight of internal structure.
- Self-weight of tunnel material.

Materials Used

In selection of a material for lining the two important aspects to be considered are strength and durability. If the tunnel material is such that it could be stable and withstand the position on its own, then the lining would only serve as a veneer.

It also serves as a protective cover against spilling or deterioration of rock due to weathering. Further the lining provides a smooth friction reducing surface.

In sub-aqueous tunnels, the lining material must be of impervious type. The various materials used for tunnel lining are described below:

Brick masonry lining: Before the advent of concrete, only bricks were used for tunnel lining. But in the modern tunnels, they are not used because of the following reasons:

- It requires a large and heavy cutting.
- The ground has to stand for long time without support.
- Unsuitable in the shield method of tunneling.

Stone masonry lining: In earlier days, only stone-masonry has been used for tunnel lining. Stone masonry has been preferred in the olden days as they are strong enough and highly durable.

Timber lining: Based on the ease availability and less cost timber linings have been used in earlier days as a temporary measure. It is not used when water proofing lining is required. It is used as a semi-permanent lining to be replaced by or supplemented with concrete lining in due course of time.

Concrete lining: Reinforced cement concrete is used as a thin lining material. In majority of the works cast-in-situ R.C.C. lining is extensively used. Steel forms of collapsible type are used which can be moved easily through the forms. The concrete mix can be shot or pumped into the collapsible steel form. In the case of invert, side wall and on arch concrete can be poured. Cast-in-situ concreting is not suitable for shield method of tunneling. In such cases, precast concrete blocks are used.

Cast-iron lining: This type of lining is particularly useful for the lining of shield driven tunnels. This type of lining has several advantages, viz., immediate strength after erection, can withstand jacking stresses, can be accurately machined to fit in the section and can be made perfectly water tight.

Structural steel lining: Structural steel is used in the form of steel ribs and linear plates. Further, it is also used in the form of welded segments of plates and of required shapes bolted together through flanges for primary lining. It is comparatively less in weight compared to cast-iron lining. If water tightness is required these linings may be welded on site. The only disadvantage is that it is subjected to corrosion and hence preventive steps have to be taken.

3.5 Piling Techniques

Piles are slender structural members normally installed by driving by hammer or by any other suitable means. Piles are usually placed in groups to provide as foundations for structures.

Piling techniques can be split into 2 categories. Displacement and replacement. In simple terms, during the displacement piling method, piles are driven into the ground pushing the ground out of the way, as we would see in sheet piling.

Displacement piling is good for contaminated sites where it costs a lot to take the spoil away. Using the replacement piling method, muck is dug out and replaced with the pile. We can use far bigger piles using replacement piling.

Displacement piling methods are typically:

- Pre-cast concrete driven piles.
- Thick wall driven steel tubes.
- Thin wall bottom driven piles.
- Timber piles.
- Screw piles.
- Helical displacement piles.

- Vibro concrete columns (drive cast insitu).

The advantages of displacement piling are:

- Self-testing as driven to refusal or "set".
- No pile arising's to dispose.
- Little disturbance.
- It is used limited access only.
- It gives high production.

The disadvantages of displacement piling are:

- Cannot penetrate obstructions.
- Cannot always penetrate clay.
- Vibration and noise may be an issue.

Replacement piling techniques are typically:

- Open hole auger piles.
- Continuous flight auger (CFA).
- Large diameter rotary piles.
- Odex piles.
- Tripod piles.

The advantages of replacement piling are:

- Effectively vibration free.
- Installed into non cohesive and water bearing soils.
- High production.
- Restricted access.

The main disadvantage of replacement piling is that:

- It produces excavated material which requires removal off site.

3.6   Well and Caiss

Caissons and well foundations are structural boxes or chambers. These are sunk in place through the ground or water by excavating below the bottom of the unit which enables the caisson to reach the final depth. These structures have a large cross sectional area and hence provide high bearing capacity, which is much larger than what may be offered by a cluster of piles.

Types of Caissons

Caissons are grouped under three categories, viz.:

- Open caissons
- Box caissons
- Pneumatic caissons

1. Open Caissons

Open caissons are concrete or masonry shafts which remain open both at the top and at the bottom during construction. It contains one or more wells for excavation called a monolith. The well foundation used in India is an open caisson with some modifications in the construction procedure.

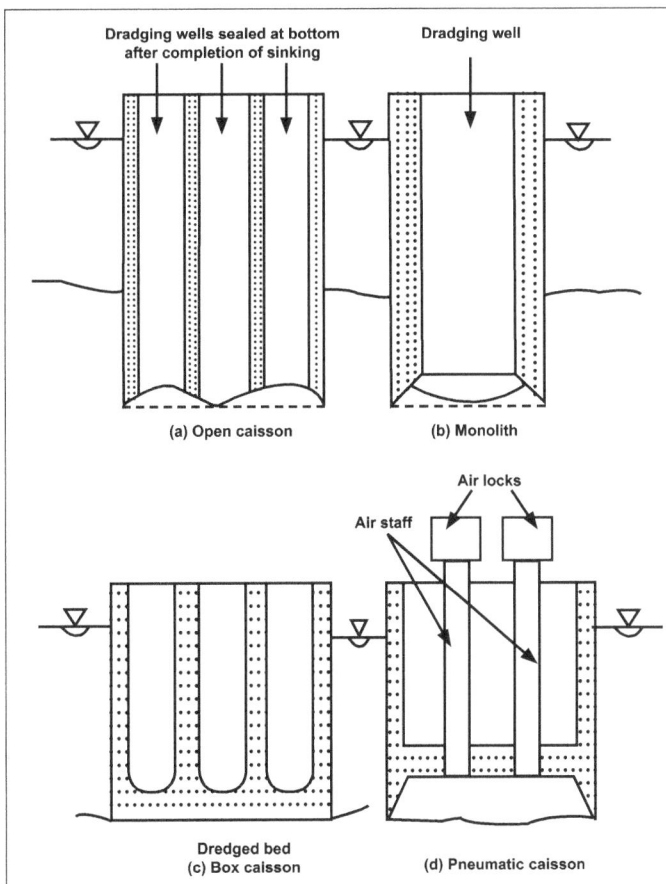

Caisson.

The caisson is sunk in place while the soil is removed from the inside. This process is continued till the well is sunk to the desired depth.

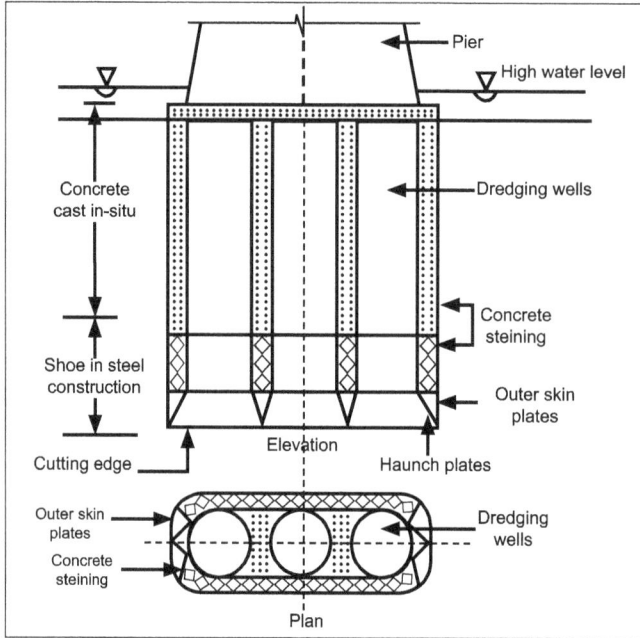

Details of an open caisson.

After this, bottom portion inside is concreted. The well is pumped dry after maturity of the bottom concrete seal and later filled with concrete or sand. If filled with sand the top is covered with concrete.

2. Box Caissons

Box caissons are closed bottom structures. They are constructed on land and transported and sunk into the prepared foundation below the water table.

3. Pneumatic Caissons

Pneumatic caissons have bottom and top are closed with an open working chamber. Compressed air is used to stop the entry of water into the working chamber. The excavation and concreting are done in a dry condition.

The caisson is sunk as the excavation is preceded. After reaching the desired depth, the working chamber is filled with concrete.

Well Foundation

Open - well - type caissons, simply called as well foundations, have been used in India for the foundation of river bridges for several years. In earlier days, the masonry of the wells was built on timber curbs.

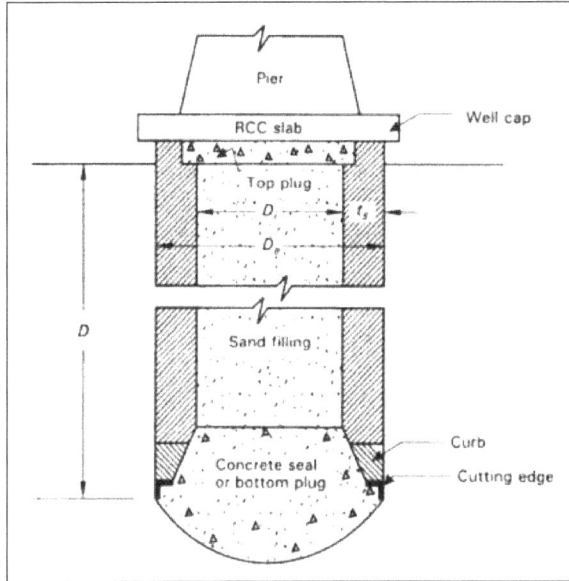

A typical section of a well foundation.

Well foundations are constructed on natural ground or over artificially formed ground. The curbs are pitched in the correct position. Then, the well is sunk into the ground to the desired level by grabbing the soil through the dredge holes formed by the masonry or concrete of steining.

A typical sectional elevation of a well foundation is shown in Figure. The bottom of a well is provided with a steel cutting edge. This tapered portion, called a well curb, is sufficiently strengthened with adequate reinforced concrete to take heavy loading.

The body of the well is called steining. Steining is made out of brick masonry, stone masonry, mass concrete or reinforced cement concrete. After sinking the well to required depth the dredge hole is cleaned and filled with cement concrete.

The bottom seal is called bottom plug. The remaining portion of the well may be filled partially or fully with saturated sand, water or left hollow. At the top of the well, another cap is provided with plain or reinforced cement concrete.

This cap is provided so as to transmit the imposed load, from piers or otherwise, uniformly to the wall of the well. A well cap is provided as the top most layer to comfortably accommodate the pier.

Types of Wells

Based on the shapes the wells are named as:

- Circular wells.

- Twin - circular wells.

- Double - D wells.

- Double octagonal wells.

- Single and double rectangular wells.

- Multiple dredged-holed wells.

Out of these, circular, twin - circular and double D are most commonly used in India.

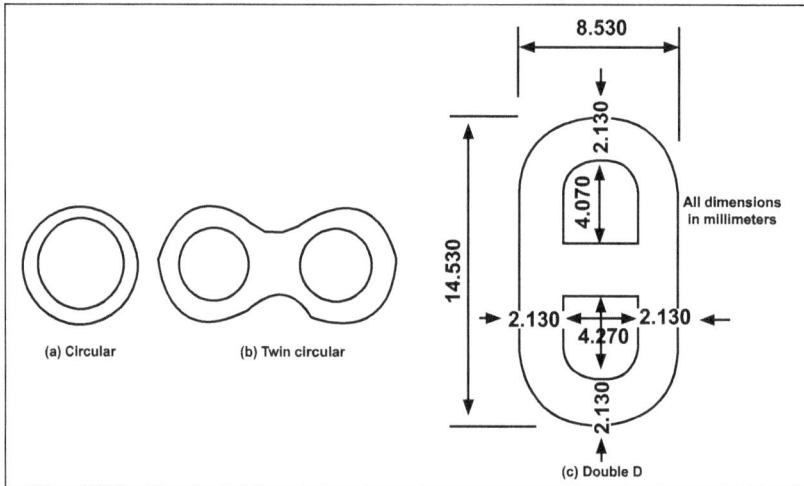

Common types of wells.

1. Circular Wells

These wells have uniform strength in all directions. The sinking process is easy because of the weight per square of peripheral surface is the highest. The construction is simple and any tilts and shifts can be rectified easily.

2. Twin - circular Wells

In such wells, two wells are placed close together and connected by a common well cap. The spacing between the wells and the diameters of the wells can be adjusted so as to accommodate the width and length of the pier. This type of well has all the advantages of a circular well. Further the well is used with advantage when the sinking depth is less and hard foundation material is available.

3. Double - D Wells

This is a commonly adopted one for deep foundations and major bridges. The length of the well is restricted to twice its width. The presence of two wide dredge holes makes the casting and sinking more effectively. Deep wells may undergo some cracks due to large bending moment.

3.7 Sinking Cofferdam

Different types of new civil engineering constructions are coming up which may require cut and cover work for public transportation, deep excavation or two or three basement floors or river diversion works, etc. These type of works require some type of temporary or permanent retaining structures which are termed as cofferdams.

Types of Cofferdams

Different types of cofferdams are as follows:

1. Cantilever Sheet Piles

Cantilever sheet piles may be used for small cofferdams. But these cofferdams are susceptible to large leakage and damage during floods.

2. Braced Cofferdams

In this type of cofferdams, the sheet piles are braced. These are economical for small to moderate height. But susceptible for damage due to floods.

3. Earth Embankments

This type of structure is more stable and can be constructed to any reasonable height.

4. Double Wall Cofferdams

In these types, sheet piles are tied by the rods and the walls are of sheet piles. These are suitable for moderate height.

5. Cellular Cofferdams

These are constructed using straight-web-sheet piling. These are suitable for moderate and large height.

Sinking Cofferdam to Bedrock Sinking

The sinking of cofferdam to the bedrock should follow the following principles:

The inclination and the displacement appeared during sinking and outgassing should be adjusted by the displacement first and then the inclination. The purpose of displacement adjustment is to make proper inclination for sinking. The inclination angle should be controlled within 1%.

To guide the mud suction and avoid sand overturning, the mud elevation within the cofferdam should be measured carefully and timely.

In the early stage of sinking, supplemental water should be pumped into the cofferdam to ensure that the water elevation inside the cofferdam is not lower than outside the cofferdam.

In sinking the overburden mud layers, the weight of cofferdam should be larger than the friction forces and the sinking coefficient should not be less than 1.25. Therefore, the underwater concrete should be poured several times to sink the cofferdam smoothly and enhance the strength of cofferdam to resist the bigger waterhead.

The elevations of riverbed should be measured periodically when sinking the cofferdam. The elevation difference around the cofferdam should be controlled within 3 m based on the location of cofferdam center. In case of larger elevation difference, protective measures such as dropping sand bags and flagstones or mud suction at higher locations should be taken.

Sinking Cofferdam.

3.8 Cable Anchoring and Grouting

Ground reinforcement includes the techniques of ground anchoring, cable bolting and rock bolting. Ground anchors tend to be longer with higher capacity and are usually associated with civil infrastructure projects.

Cable bolts are used in stability problems that lie between the two and are commonly used in mining engineering.

Anchors in Rocks

In the majority of moderately weak to strong rocks, rotary or rotary percussive open hole drilling with air flush, followed by normal grouting techniques will achieve the required grout/rock bond capacity.

When fissures or voids are detected by loss of flush, by water ingress, by water testing, or inability to maintain a head of grout within the bore, then pre-grouting operations or alternatively pressure grouting operations may be required.

Normally cement grouts are injected but if fissures are known to be wide, sanded mixes may be used.

In coarse grained weak rocks similar techniques or alternatively rotary water flush drilling can be used and in most conditions a reasonable anchorage capacity can be obtained.

Anchors in Clay

In order to enhance the capacity of the anchorage within the normal range of fixed lengths, either under reaming or soil fracturing systems has been employed. More recently, the single bore multiple anchor system has been allowed efficient use of non-enhanced bore holes and attained loads of 3500kN.

The fracturing of soil prior to tendon installation, involves a larger diameter steel manchette, which after treatment remains in-situ. Treatment may be carried out over a 2 or 3 day period prior to tendon installation, by repeatedly injecting grout through manchette valves at 1/2 m centers in the fixed length.

The anchor tendon is then, after pre grouting treatment, installed and grouted within the large tube. The tube must efficiently transfer the entire load from anchor tendon and internal grout to the external grout and then into the ground.

Anchors in Granular Materials

Anchors are installed in granular deposits by drive drilling with a knock-off bit or by use of duplex drilling techniques. Drive drilling involves the percussion driving of a strong casing with a conical lead bit resulting lateral soil replacement and no flush recovery.

The lead bit is knocked off the casing allowing tendon installation and pressure grouting during withdrawal. There are limitations in the depth penetrable. Duplex drilling involves the advancement of both drill rods and drill casing, utilizing casing sizes of the 80 to 150mm ranges.

Either air or water flush or augers can be used, although bit wear and casing wear may well be considerably higher without lubrications and cooling by water.

Grouting

Grouting is a process whereby stabilizers, either in the form of a suspension or solution, are injected into sub–surface soil or rock for one or more applications.

Basic Functions of Grouting

The three basic functions of grouting are:

- Permeation or penetration: The grout flows freely with minimal effort into the soil voids or rock seams.

- Compaction or controlled displacement: In this condition, the ground remains more or less intact as a mass and exerts pressure on the soil or rock.

- Hydraulic fracturing: Hydraulic fracturing or uncontrolled displacement occurs when the grouting pressures are greater than the tensile strength of the soil or rock being grouted. Then, the latter materials fail and the grout rapidly penetrates into the fracture zone.

1. Suspension Grouts

Suspension grouts consist of solid particles like soil, cement, lime, asphalt emulsion, etc., carried in water. The solution grouts are numerous, viz., aqueous, non–aqueous, colloids, etc. Particles in a suspension grout are of silt size and hence these materials cannot be rejected into the pores of soils finer than medium to coarse sand size.

For successful grouting of soils, the groutability ratio should be greater than 20. This criterion basically limits the use of suspension grouting to permeation of sands and gravels.

Other considerations that must be taken into account in grouting design are the grout's setting time and its stability. Grouting can be done with soil, soil–cement mixers, cement or lime.

2. Solution Grouts

Solution grouting is done using "one-shot" or "two-shot" systems. In the one-shot method, all required chemicals are injected together after re-mixing. Setting times are controlled by varying the catalyst concentration according to the grout concentration, water composition and temperature.

The two-shot method, where in one chemical is injected followed by injection of a second chemical which reacts with the first to produce a precipitate in the soil pores, may also be used. Two-shot systems are slower and require higher injection pressures and more closely spaced grout holes.

3.9   Driving Diaphragm Walls and Sheet Piles

Diaphragm Wall Construction

Diaphragm wall is a continuous wall constructed in ground to facilitate certain construction activities. It is used as a:

- Retaining wall construction.

- Cut-off provision to support deep excavation.

- Final wall for basement or other underground structure (e.g. tunnel and shaft).

- Separating structure between major underground facilities.

- Form of foundation (barrette pile –rectangular pile).

Diaphragm Wall

Diaphragm wall is a reinforced concrete structure constructed in-situ panel by panel. The wall is usually designed to reach very great depth, sometimes up to 50 m, mechanical excavating method is thus employed.

Typical sequence of work includes:

- Construct the guide wall.

- Excavation to form the diaphragm wall trench.

- Support the trench cutting using bentonite slurry.

- Inert reinforcement and placing of concrete to form the wall panel.

Further Explanation on the Work Sequences

Guide wall is two parallel concrete beams constructed along the side of the wall as a guide to the clamshell which is used for the excavation of the diaphragm wall trenches. Trench excavation in normal soil condition excavation is done using a clamshell or grab suspended by cables to a crane. The grab can easily cut through soft ground. In case of encountering boulders, a gravity hammer (chisel) will be used to break the rock and then take the spoil out using the grab.

Excavation support of inside the trench cut can collapse easily. Bentonite slurry is used to protect the sides of the soil. Bentonite is a specially selected fine clay, when added to water, forms an impervious cake like slurry with very large viscosity. The slurry will produce a great lateral pressure sufficient enough to retain the vertical soil.

Reinforcement is inserted in the form of a steel cage, but may be required to lap a few sections in order to reach the required length. Placing of concrete is done using tremie pipes to avoid the segregation of concrete. As concrete being poured down, bontonite will be displaced due to its lower density than concrete. Bontonite is then collected and reused.

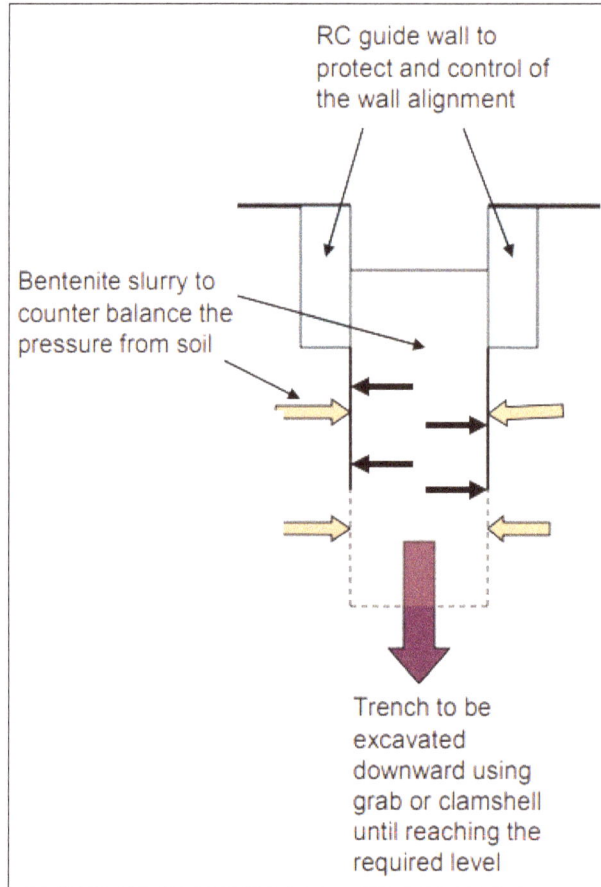

RC guide wall to protect and control of the wall alignment

Bentenite slurry to counter balance the pressure from soil

Trench to be excavated downward using grab or clamshell until reaching the required level

Diaphragm wall.

Joining for the Diaphragm Wall Panel

Diaphragm wall cannot be constructed continually for a very long section due to limitation and the size of the mechanical plant. The wall is usually constructed in the alternative section. Two stop end tubes will be placed at the ends of the excavated trench before concreting.

The tubes are withdrawn at the same time of concreting, so that a semi-circular end section is formed. Wall sections are formed alternatively leaving an intermediate section in between. The in-between sections are built similarly, afterward but without the end tube. At the end a continual diaphragm, wall is constructed in the panel sections tightly joined by the semi-circular groove.

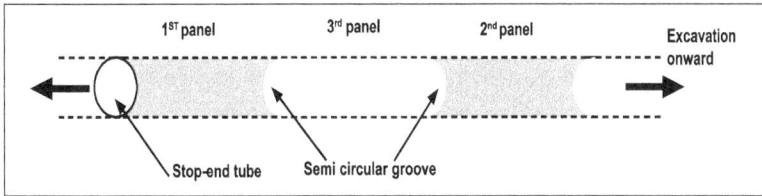

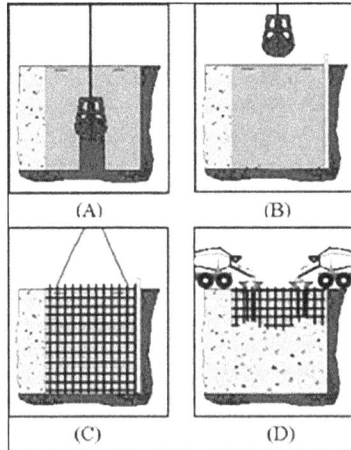

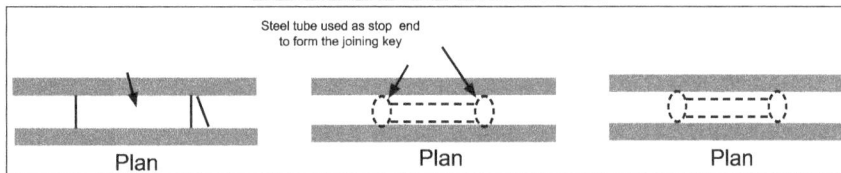

Joining of Diaphragm wall.

Construction process for a diaphragm wall panel:

- Forming the guide wall and using it in the trenching operation.
- Clamshell to excavate the trench to form a diaphragm wall panel.
- Fixing and placing of reinforcing cage.

Application of diaphragm wall:

- Commonly used in congested areas.
- Can be installed in close proximity to existing structure.
- Practically suited for deep basements.
- Used in conjunction with "Top Down" construction technique.

Sheet Piles Structures

Materials used for sheet pile structures range from simple wooden planks, light gauge sheet metals and heavy sections made of reinforced concrete and structural steel members.

Wooden Sheet Piles

Wooden planks of 5 to 30 cm or still thicker materials are used in wooden sheet piles. The simplest form is single planks which are driven edge to edge. In double–planks type, they are nailed together in staggered position, thus forming a lap at each joint (figure a).

Three–plank type is used in which the central plank is of size 3 to 6 cm thick. Tongue and groove types are also used. If the planks are impregnated with preservatives and mopped with preservatives on sawed or cut surfaces, they can be used as a permanent structure provided they are permanently under water.

In order to ease driving the bottom of the pile may be cut with a level and provided with a metal driving shoe.

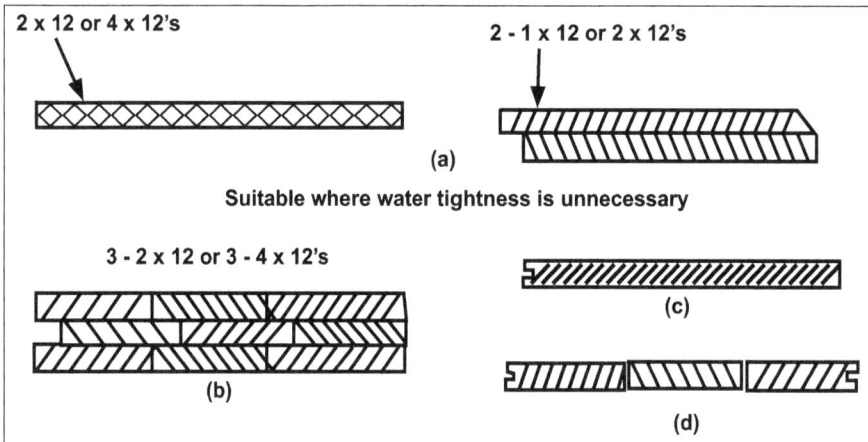

Wooden sheet piles (a) planks (b) Wakefield (c) tongue and groove (tongue and groove cut in the mill) (d) splined (grooves cut in the mill, splines driven after piles are in place).

Concrete Sheet Piles

Concrete sheet piles are precast members which are made to withstand handling stresses during construction and permanent stresses during service. Short piles are handled by lifting at one point whereas for long piles two or more pick up points may be necessary to reduce the handling stresses.

In order to avoid cracks due to shrinkage and handling a certain minimum amount of reinforcing is desirable. The reinforcements should be placed at closer spacing at the top and bottom of the pile to reduce the possibility of damage due to driving.

Concrete sheet piles are grouted for water tightness. In such cases expansion joints above the dredge line should be allowed. As the concrete piles are heavy and bulky, they require heavy equipment to drive and handle. Further as they are voluminous, they displace more soil and have high frictional resistance. These piles are also subjected to greater bending stresses than other type of sheet pile walls.

Steel Sheet Piles

Steel sheet piles are rolled structural members. Each sheet pile is provided with inter-locks so as to engage with one another. Based on the manufacturing concern the sheet pile pattern varies along with interlocking arrangements. A few types of sheet piles are shown in Figure.

In order to resist large bending moments the arch and the Z-piles are used. To resist larger bending moments the deep–arch web and Z-piles are used. The shallow–arch piles or smaller section modulus can be used where lesser bending moments are met with. Straight web–sheetings are used where the web is required to take tension (in case of cellular cofferdam).

In the case of cofferdams, the section may be made up or formed into certain standard joints, such as T_s, Y^s and crosses. Straight web piles are used where piles are subjected to tension and interlocking strength is of primary importance–primarily in cellular cofferdams.

Finger and thumb type interlocks provide water tightness and strength. Arch web piles are used to resist large bending–cantilever and anchored walls. They are suitable where difficult driving are contemplated and have finger and thumb type interlocks. Z-piles have the highest bending strength. Ball and socket type interlocks offer least driving resistance.

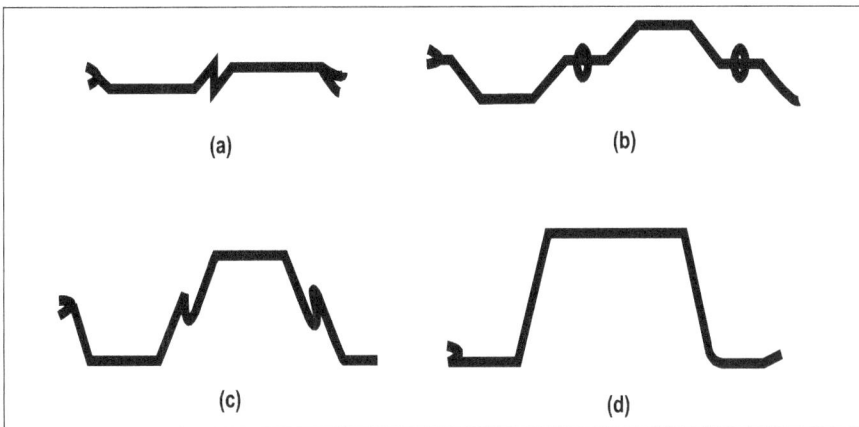

Steel sheet piles (a) straight web piles (b) shallow arch piles (c) deep arch piles (d) Z-piles.

Advantages of Steel Sheet Piles

Steel sheet piles have several advantages and some of them are given below:

- It can resist hard driving – stresses.

- It is of light weight.

- It can be reused several times.

- It has long life both above and below water table.

- During construction or afterwards the pile length can be increased either by welding or bolting.

- Joints are less susceptible to deform when wedged in hard soils.

Construction of Sheet Pile Walls

Sheet piles are driven in place of installation using the same type of equipment used for driving bearing piles. Sheet piles are driven with the tongue or ball in leading position. Sheet-piles are driven in pairs. This is done for faster operation and better economy.

Further, driving a pair of piles removes the additional energy required to overcome the inter lock friction between the pair if they are driven separately. Among the type of piles, the arch-web type has greater rigidity during driving. In order to maintain proper alignment of sheet piles, guides or temporary wailing are used. These guides are heavy horizontal timbers or steel beams supported by stakes.

Sheet piles tend to lean outward also tend to creep along the plane of the wall and in the direction of piling. Such difficulties can be overcome by driving in panels. That is a panel consisting of a few piles are driven to part penetration at a time. After driving a few number of panels, the first few panels of piles are driven to the required depth.

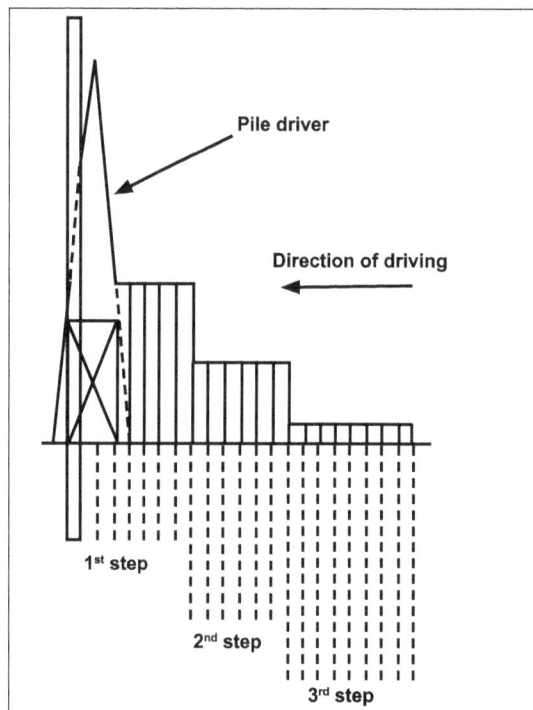

Driving sheet piles.

3.10 Shoring for Deep Cutting

Shoring is the means of providing support to get stability of a structure temporarily under certain circumstances during construction, repair or alteration. Such a circumstance may arise when:

- The stability of a structure is endangered due to removal of a defective portion of the structure.

- The stability of a structure is endangered due to the unequal settlement during construction itself or in long run.

- Certain alterations are required to be done in the present structure itself, e.g., re-modeling of walls, changing positions of windows, etc.

- Alterations are carried out in adjacent building for re-modeling, strengthening of foundations, etc.

Types of Shoring

Shores are classified under the following categories:

- Raking or inclined shores.

- Flying or horizontal shores.

- Dead or vertical shores.

1. Raking or Inclined Shores

In this type of shoring, inclined members are adopted to provide temporary support to the external walls from the ground. These inclined members are called as rakers. A raking shore primarily consist of rakers, needles, cleats, braces, wall plates and sole plates.

The wall plate is secured against the wall by means of square needles. These needles penetrate into the wall for a depth of 15 cm and prevent the wall platform from sliding against the wall while the wall plate distributes the pressure evenly.

Further, the needles themselves are strengthened by providing wooden cleats. Timber braces are used to interconnect the inclined rakers. The feet of the rakers are tied together by braces and hoop iron. These are in turn connected to sole plate by means of iron dogs or dog spikes.

The details shown in Figure is a simple raking shore for a height of buildings up to 10 m. Special rakers have to be provided for multi-storeyed buildings and buildings on road side. Dealing with road side buildings the bye-laws of the local area should be followed.

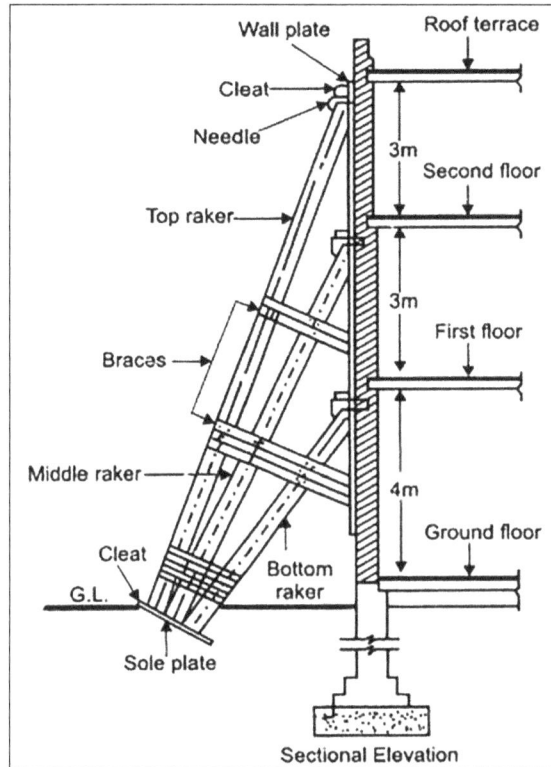

Raking or Inclined shoring for a building up to 10 m height.

2. Flying or Horizontal Shoring

In this type of shoring, horizontal supports are provided to parallel walls which have become unsafe due to some reason. This arrangement is called a simple flying shore.

This consists of wall plate, needles, cleats, struts, straining pieces and folding wedges. Like inclined shores, in this system also the wall plates are secured against the wall by means of needles and cleats.

The horizontal shore is kept in the required position by means of wedges, needles and cleats to the wall plate. The inclined struts are supported by the needles at their one end straining at the other end. In turn, the straining is fixed to the horizontal shore.

This type of single flying shoring can be adopted for a maximum distance of about 9 m between the adjacent parallel walls. When the distance is between 9 to 12 m, a compound or double flying shore is provided. It is to be noted that both the horizontal shores are symmetrically placed with respect to the floor levels.

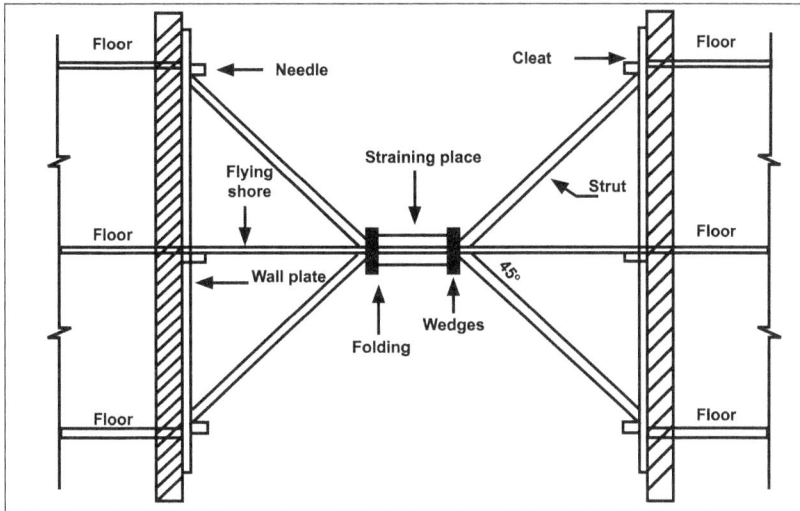

Details of single flying shore.

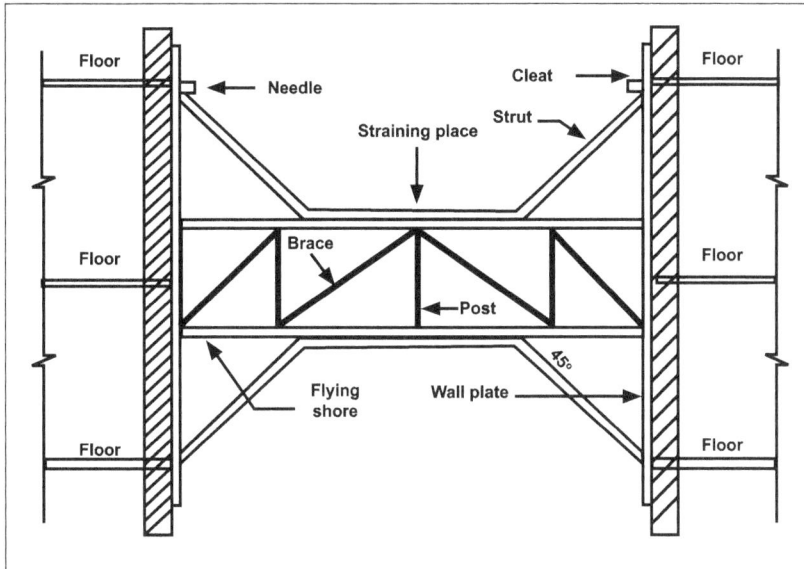

Details of double flying shore.

3. Dead or Vertical Shoring

These shores are placed vertically and are used for temporarily supporting the wall while the lower part of which are to be removed for repairs. By this arrangement, the whole load of the roof is supported by these shores.

Needles are used to transfer the load from the walls. These are first inserted into small wall opening. Horizontal beams are laid along the floors. This beam supports the dead or vertical shore and distributes the load evenly on the base. The dead shores are installed in between the beams and the needles by means of wedges.

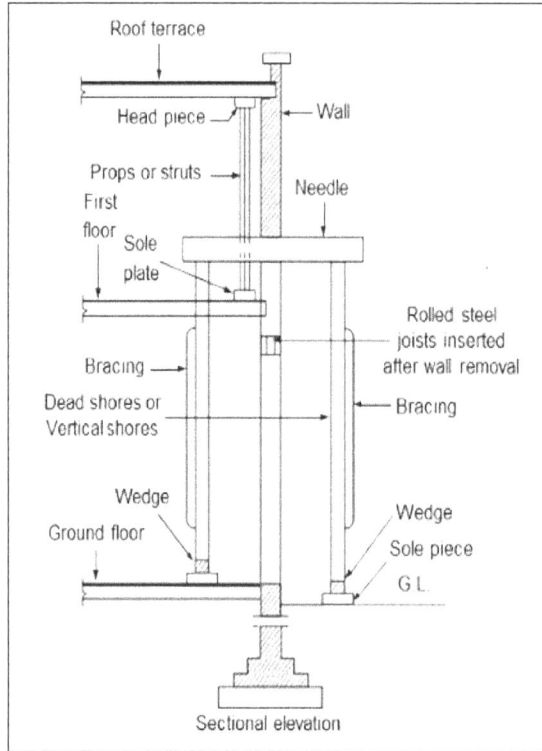

Use of dead-shores for repairs.

In order to avoid any damage, the usual walls are supported with props before removing the desired portions. In the same way, the windows or other openings are duly strutted.

SI. No	Type of Structural Member	Period
1.	Vertical side supports of slabs, beams columns and walls	2 days
2.	Slabs with vertical supports or props left under	7 days
3.	Soffits of beams with props	7 days
4.	Bottom of slabs up to a span of 4.5 m	7 days
5.	Bottom of slabs above 4.5 m span	14 days
6.	Bottom of beam up to 6 m span	14 days
7.	Bottom of arch ribs up to 6 m span	14 days
8.	Bottom of beams over 6 m span	21 days
9.	Bottom of arch ribs over 6 m span	21 days

3.11 Well Points

Filter wells or well points are small well screens of sizes 50 to 80 mm in diameter and 0.3 to 1 m length. Well points are made of brass or stainless steel screens and of closed ends or self-jetting types. When well points are required to remain in the ground for a long period disposable plastic well points are used.

The plastic well-points are of nylon mesh screens surrounded by flexible plastic riser pipes. Water drawn through the screen enters the space between the gauze and the outside of the riser pipe through holes drilled in the bottom of the pipe and then reaches the surface. The well-points are installed by jetting them into the ground.

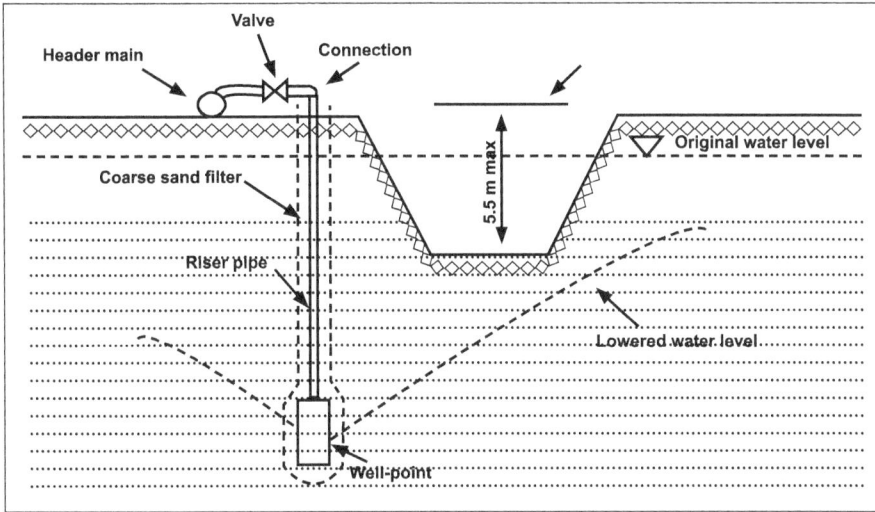

Single stage well-point installation.

Ten liters/m is the capacity of a single well-point with a 50 mm riser. Spacing of well points is based on the permeability of the soil and the time taken for the required draw down. In general, the satisfactory spacing is 0.75 to 1m in fine to medium dense sand.

A spacing of 1.5 m may be necessary in a fairly low permeability soils such as silty sands. A spacing of 0.3 m is sufficient in highly permeable soils such as coarse gravels.

A well-point system comprises of well-points which are attached to riser pipes, which extend a short distance above the surface of the ground, where they are connected to a large pipe called header.

The header pipe is connected to the suction of a centrifugal pump. A well-point system may include a few or several hundred well-points (generally 50 to 60 well-points) all connected to one or more headers and pumps.

Well-point systems are very effective in solving subsurface water problem on construction sites. These systems are used to provide dry work areas below the water table for the following works:

- Foundation work including buildings, bridges, dams and dry dock.

- Trench work including buildings, bridges, dams and dry dock.

- Tunnel work such as subway construction.

The advantages of well-point systems are as follows:

- This system is adoptable in many jobs, as it does not interfere in the working of the excavation-hauling equipment and the job.

- This system actually stabilizes the soil in the work area.

- It minimizes the possibility of settling and damage to adjacent structures.

- It prevents the bottom of excavation from heaving under excessive hydrostatic pressure.

- It prevents slope failure or sloughing.

- When unexpected de watering conditions develop well-point system can be readily adopted to it.

The serious limitation of well-point system is the suction lift. A lowering of about 6 m below pump level is generally possible beyond which excessive air shall be drawn into the system through joints in the pipes, valves, etc., resulting in loss of pumping efficiency. If the ground consists mainly of large gravel, stiff clay or soil containing cobbles or boulders, it is not possible to install well-points.

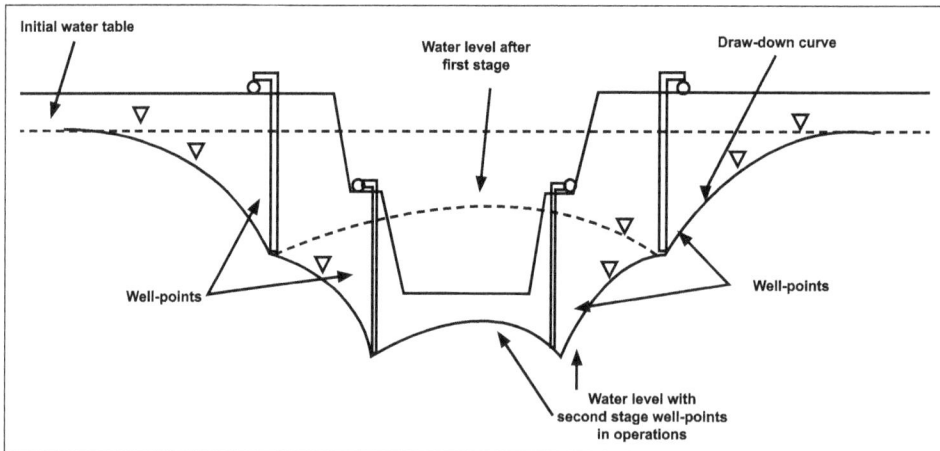

Multistage well-point operation.

For dewatering deeper excavations the well-points must be installed in two or more stages as shown in figure. There is no limit to the depth of draw-down in their way, but the overall width of excavation at ground level becomes very large. On the other hand it is possible to avoid multi well-point stages by excavating down to water table level before installing the pump and header.

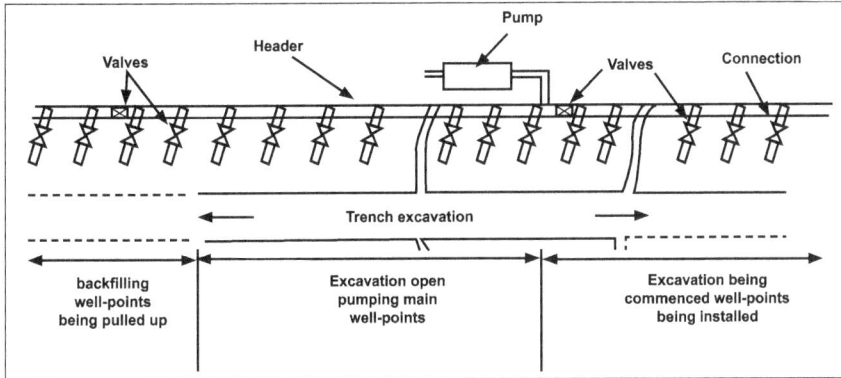

Single stage well-point installation by progressive system.

Well-points are installed either by the progressive system or the ring system. In the progressive system the header is laid out along the sides of the excavation.

Pumping is done continuously in one length as further points are jetted ahead of the pumped down section and pulled out from the completed and backfilled lengths.

Only one header is sufficient for narrow excavations. But for wide excavations or stratified soils of low permeability, the header must be placed on both sides of the excavation.

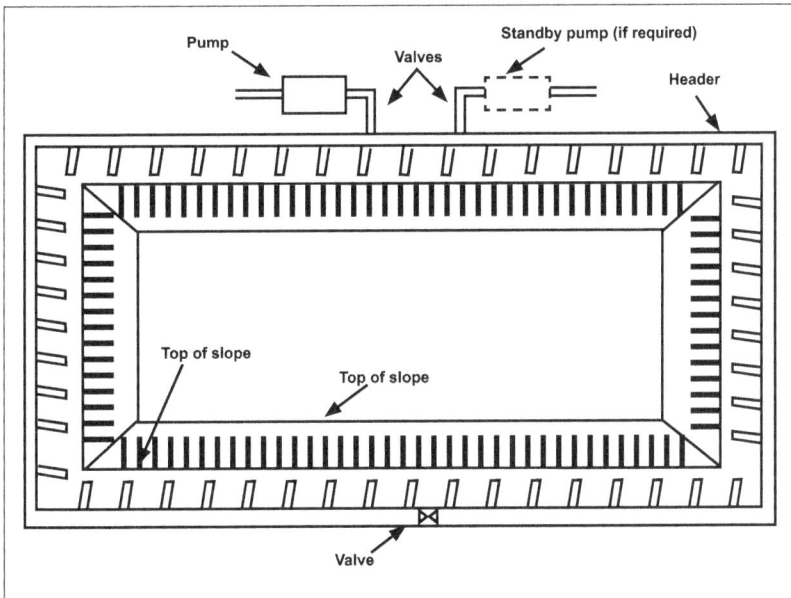

Well-point installation by the ring system.

In the ring system of installation (Figure) the header main is placed completely to surround the excavation. The progressive system is suitable for trench work whereas the ring system is most suitable for rectangular excavations such as for piers or basements and for long trenches.

3.12 De-Watering and Stand by Plant Equipment for Underground Open Excavation

Vacuum Dewatering System

Gravity methods explained so far are not very effective in fine-grained soils. Such soils can be stabilized by means of a vacuum well or well-point system. A vacuum de-watering system primarily consists of well or well-points with the screen and riser pipe.

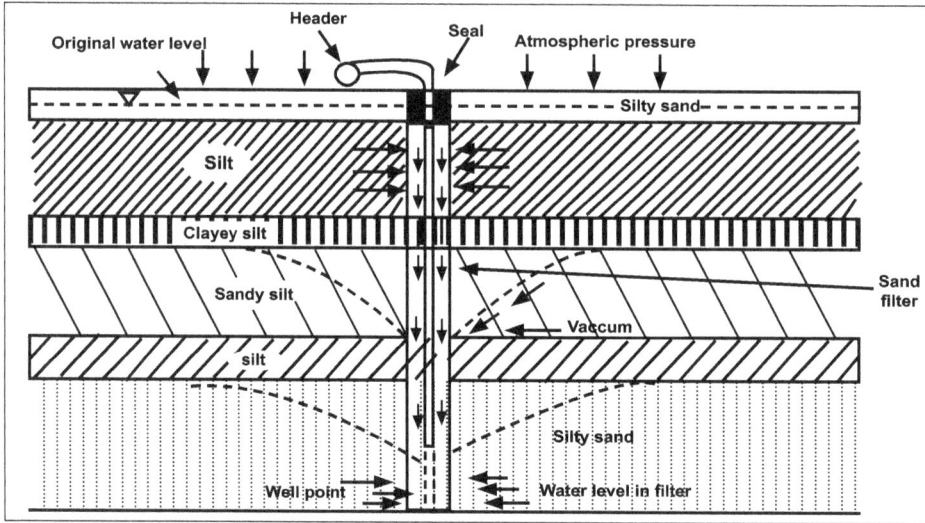

Vacuum dewatering system.

A stabilizing fine soil such as bentonite or impervious soil seal is provided at the remaining portion of the hole. By creating and maintaining a vacuum in the well screen and the sand filter, the flow towards the well is increased. A closer spacing is required for proper de-watering.

Dewatering by Electro-osmosis

This is also a method applicable for fine grained soils. This is not a general pumping method but collecting the water through some process to a well and pumping out.

If the vacuum well-point or well-point system is ineffective, application of an electrical gradient may be made. In a fine grained soil stratum, when a direct electric current is passed through a saturated soil stratum, water moves towards the cathode.

If the water is removed at the cathode, the soil decreases in volume resulting in increased shear strength. This process is called de-watering by electro osmosis.

Electro-osmotic flow is dependent on the porosity of the soil and the electrical potential. A comparison is made in Figure between the electro-osmotic flow with hydraulic flow through a single capillary.

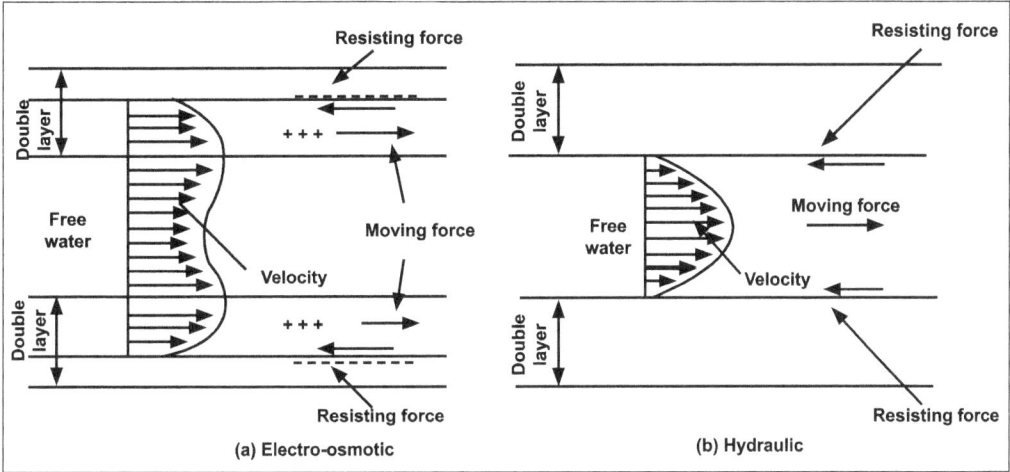

Comparison of electro-osmotic and hydraulic flows.

The general layout of the electrode is based on the purpose for which they are intended. Electrode arrangements for two field situations are shown in Figure. Sheet piles of any shape and old pipes of 25 to 50 mm diameter can be made as anodes.

Since the anodes corrode considerably in the course of a few weeks of electro-osmotic treatment, they should be replaced as soon as the current drops to less than 30% of the initial consumption. For cathodes perforated tubes are used and the cathode wells are connected to a pumping system. Electro-osmotic method is used only when other methods fail as the cost of installation and maintenance are very high.

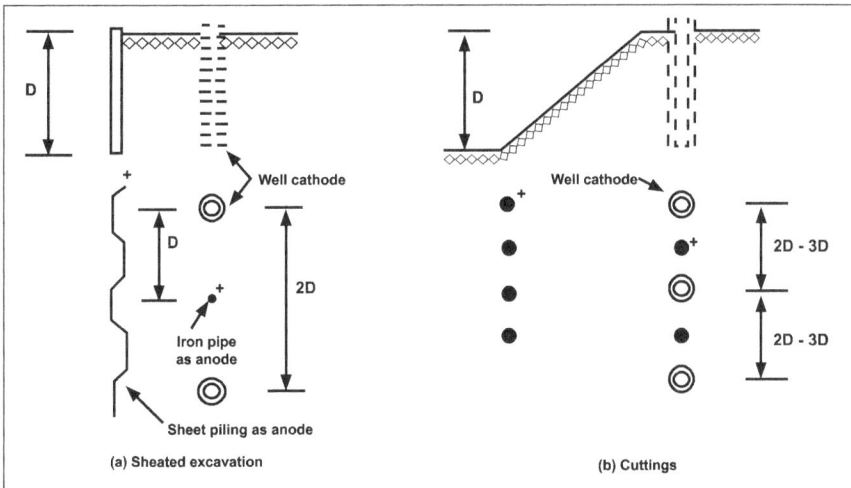

Electrode arrangements.

Freezing Method

Freezing of pore water in soil is the most effective method of thermal stabilization.

The main features of the system are:

- Around the excavation a ring or rectangle of bore holes at 1 to 1.5 m spacing are sunk.

- Lining with 100 to 150 mm steel tubing with closed bottom is made in bores.

- Inner tubing of 38 to 75 mm diameter with bottom open is inserted.

- Tops of inner tubes are connected to the main ring carrying chilled brine from refrigeration plant.

- The brine is pumped down the inner tubes.

- The brine rises up in the annular space between the inner tube and outer casing and returns to the refrigeration plant connected to the main ring.

It takes two to four months to get the strata frozen. In order to find the state of freezing a bore hole lined with perforated pipes is drilled near the center of the area. This pipe acts as a tell-tale.

As the formation of ice starts the ground is compressed and water is expelled from the strata. The rising of water in tell-tale pipe indicates the freezing of the strata. Then the excavation can be started.

Disadvantages of the system:

- As the bore holes have to be sunk at close spacing the cost is high.

- It takes long time to commence the project.

- There is some difficulty in operating compressed air tools in low temperature.

Super-Structure Construction

4.1 Launching Girders, Bridge Decks and Off Shore Platforms

A bridge is a structure constructed to provide passage for a road or railway over an obstacle such as river, valley, etc., without disturbing or closing the way beneath.

The various components of a bridge are:

- Foundations.
- Piers.
- Abutments.
- Bank connections.
- Approaches.
- Bearings.
- Decks.
- Hand rails.

Components (i) to (v) are grouped under substructure and (vi) to (viii) as superstructure of a bridge.

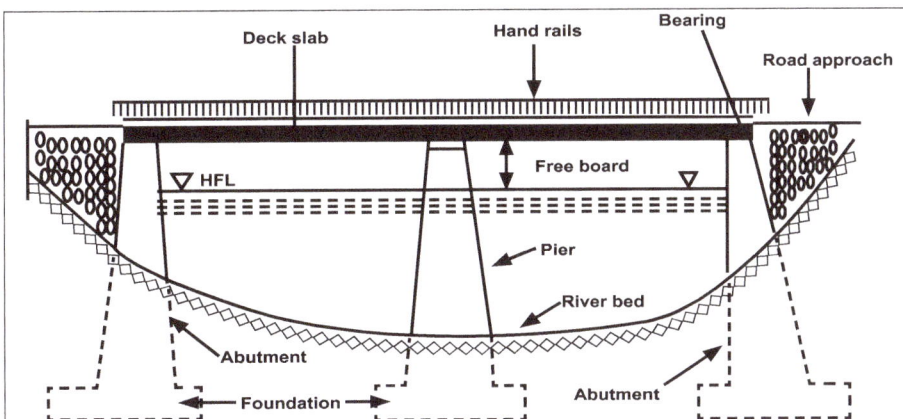

Superstructure of bridge.

The basic components of a superstructure of a bridge may be considered as decks, bearings and hand rails. A deck is the top portion of a bridge provided to form roadway or railway. It spans continuously from abutment to abutment over the piers. It may consist of a slab, girders, trusses, arches, etc.

Bearings are part of a bridge provided to transmit the load from the superstructure to the substructure in such a manner that the bearing stresses induced in the substructure are within permissible limits and also to allow for linear and angular movements.

Handrails, parapet or guard stones are provided on either side of the deck along the roadway to safeguard the moving vehicles and the passengers on the bridge deck.

Reinforced Cement Concrete Bridges

Reinforced cement concrete is well suited for the construction of bridges in the small and medium span ranges. Different types of R.C.C. bridges which are usually constructed are:

- Slab bridges.

- T-beam (girder and slab) bridges.

- Hollow girder bridges.

- Balanced cantilever bridges.

- Continuous girder bridges.

- Rigid frame bridges.

- Arch bridges.

- Bowstring girder bridges.

Bridge Decks

Bridges are designed for movement of loads and personnel which need a flat surface for movement. This flat surface is called the deck surface. The deck may consist of a slab, a beam and a slab, a grillage, a box girder, multi beam, etc.

The simplest form of concrete deck is a solid slab which may be reinforced or prestressed. The weight of the deck increases with the increase of span.

The material in the middle depth of the cross section could be removed. This has lead to the formation of voided slab. The figure shows different types of bridge decks.

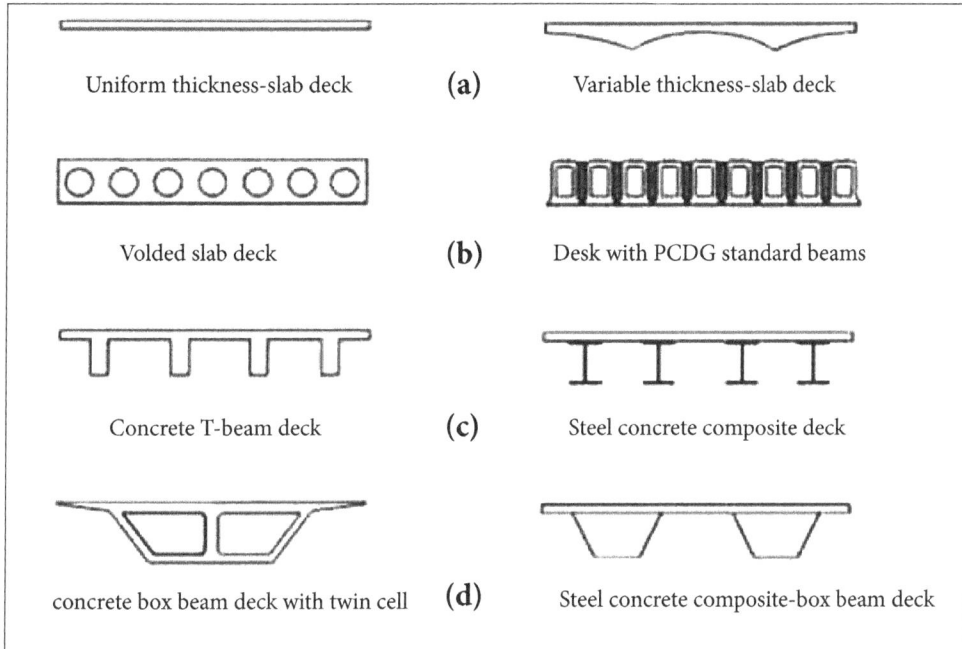

Uniform thickness-slab deck	**(a)**	Variable thickness-slab deck
Volded slab deck	**(b)**	Desk with PCDG standard beams
Concrete T-beam deck	**(c)**	Steel concrete composite deck
concrete box beam deck with twin cell	**(d)**	Steel concrete composite-box beam deck

Types of concrete bridge deck.

1. Slab Decks

Slab decks behave like a flat plate. It is a structural continuum for transferring moments, shears and torsion in all directions in the plane of the plate. The deformation of the slab depends on the support conditions.

Conventionally, in a bridge two sides are supported on bearings over the piers and the other two sides will be either free or stiffened by edge beams corresponding to elastic supports. For spans up to 10 m concrete slab decks are used.

2. Voided Slab Decks

In general, the flexural behaviour of a slab is such that the middle part of the slab will not be subjected to any stress, but only its weight to the structure.

Further, the weight increases with increase in span. In order to reduce the weight and lighten the structure, voids of cylindrical or rectangular shapes are introduced at the middle height of the cross-section.

The variation of stiffness is very small if the depth and width of the voids are less than 60%. Such an arrangement makes the deck to behave effectively as a plate.

3. Beam and Slab Decks

Beam and slab deck comprises of number of a longitudinal beams connected at the top of a slab. These beams may be connected transversely by a diaphragm or cross girder which provides stiffness for the deck.

The deck may consist of:

- A prestressed concrete beam or R.C.C beam with cast in-situ reinforced concrete slab on top.

- A steel-beam with cast in-situ reinforced concrete slab on top.

In this type of decks, the beams may be packed side by side or the beams are situated reasonably far away from each other. This system is simple and faster for construction which may be adopted for a span up to 25 m.

4. Box Girder Decks

When the space and widths increase the bottom of the beam and slab are to be tied at the bottom to keep the geometry. This has lead to evolution of Box Girders.

Eccentric loads cause higher torsional forces which could be better counteracted by a box section. This is chosen for large spans with wide decks.

Based on the type of construction method, it could be used up to 150 m span. In this type, cantilever construction is one of the regularly adopted and accepted methods of construction.

Offshore Platforms

Offshore platforms are structures constructed on the ocean to explore or to produce oil and gas from the sources found below the sea. Offshore platforms are in steel or in concrete.

Offshore platform.

Offshore platforms are used for drilling or production. Equipment reliability is of paramount importance in keeping the operation on schedule.

When equipment goes down, the ramifications are far greater than with its onshore counterparts. The equipment maintenance or the replacement of damaged parts is very costly and can involve substantial delays in getting the parts out to remote locations.

The types of offshore platforms are:

- Converted jackup barges.
- Fixed tower structures.
- Tension leg platforms (TLP's).
- Stationary floating (SPAR's).

The uses of offshore platform are:

- Connect the offshore pipeline grid.
- Provide an efficient means to platform maintenance.
- Locate compression, separation, production handling and other facilities.
- Conduct drilling operations during the initial development phase of an oil and Natural gas property.

4.2   Special Forms for Shells

A shell structure is a curved surface structure. It is a relatively thin slab which is curved in one or both directions. It is often stiffened along its edges to maintain its curvature. As shell can cover large spans without interruption of columns, shell roof is becoming very popular for industrial buildings, research laboratories, hangers and other large span buildings. Further, it has a special advantage that there is an appreciable reduction of dead weight.

Types of Shell Structures

Based on the geometry of the middle surface, shells may be classified as:

- Domes.
- Shell barrel vaults.
- Translation shells.
- Ruled surface shells.

Domes

A dome is a type of roof of semi-spherical or semi-elliptical shape. The materials used for construction of domes are stone, brick or concrete. They are supported on circular or polygon shaped walls. Domes are preferred for covering large areas and of architectural importance, such as assembly halls, gymnasiums, field houses and other monumental structures.

The domes can be either:

- Smooth shell domes.

- Ribbed domes.

Smooth shell domes may have varied or constant thickness. A lantern may or may not be provided. The dome surface is subdivided into a number of triangles by ribs.

Here, a tension ring constructed on the perimeter of the structure is a thrust resisting member. The ring is usually supported on columns spaced around the perimeter and braced to provide lateral stability for the structure.

In order to support, bearing walls are also constructed. In order to span, the space between the ribs and to support the roof deck purlins are provided. Figures show some types of domes:

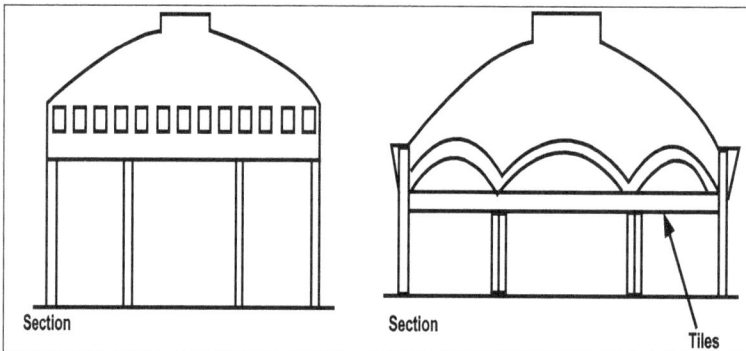

A typical spherical dome.

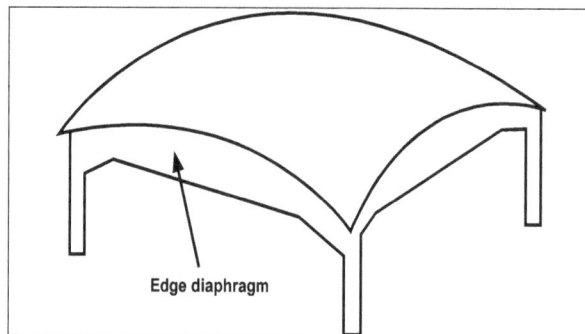

Square dome.

Shell Barrel Vault

Elements of a shell barrel vault consists of edge beams, end frame and curved membrane. Different parts are shown in figure:

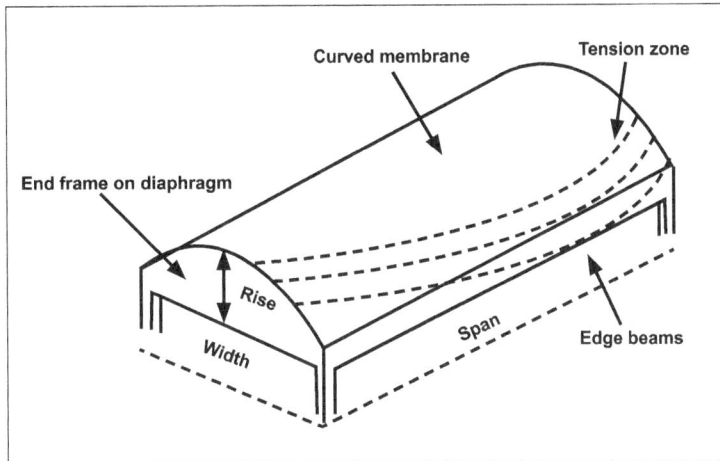

Elements of a shell barrel vault.

Translation Shells

This type of shell is developed by moving a vertical curve parallel to itself along another vertical curve. When the sliding curve is a straight line, then a cylindrical shell is developed.

Hyperbolic parabola is developed by sliding a vertical parabola with upward curvature on another parabola with downward curvature in a plane at right angles to the plane of the first. This surface is usually referred to as a saddle surface.

These surfaces have small rises so as to have flat roofs. As per geometry, such roof structures, if cut by planes parallel to the two parabolas, the edges will be parabolic and the supporting structure must be parabolic.

Ruled Surface Shells

This type of shell is developed by moving a straight line so that it ends lie on two fixed vertical curves. These vertical curves may be different types or of similar types.

If two of these curves are of similar type, then the resulting shell is a cylindrical shell. Instead, if one of the curves is circular, elliptical, etc., then the resulting shell is a conoid. Such shells also have two opposite curvatures and have saddled surfaces.

A cone is a special form of conoid in which the horizontal line is a point and the other curve is circular. Figure shows different types of doubly curved shells.

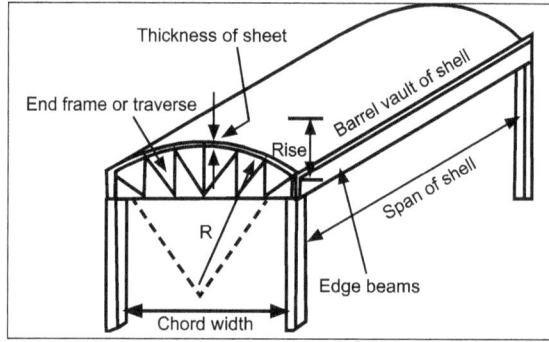

Elements of a cylindrical shell (Single Barrel).

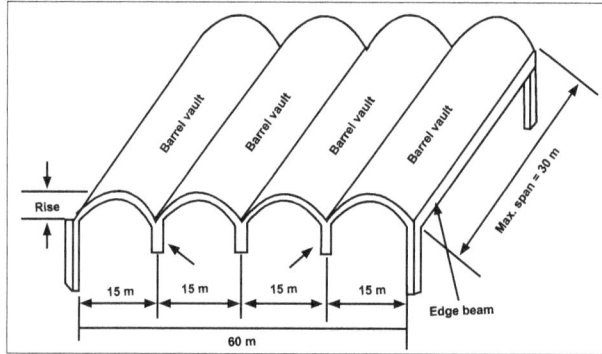

Multiple barrel type shell.

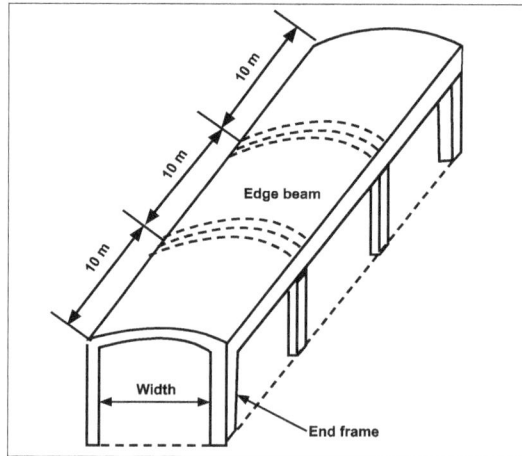

Continuous barrel type shell.

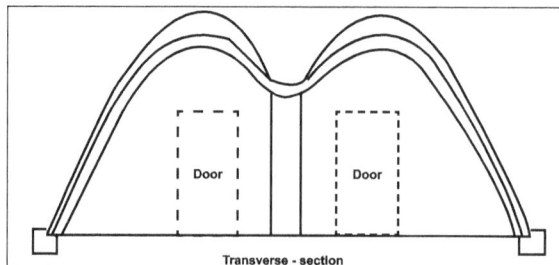

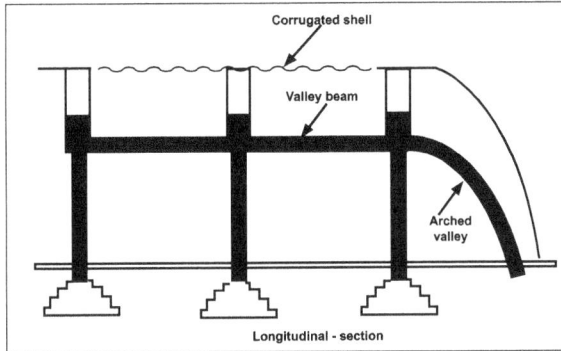

Section of Shell.

2. Merits and Demerits of Shell Roofs

Merits

- Structural shell is capable of transmitting loads in more than two directions to support.
- Shells are structurally highly efficient when they are so shaped, proportioned and supported.
- Shells transmit the loads without bending or twisting.
- Shells have relatively small thickness compared to other dimensions.
- Shells provide uninterrupted space without columns.

Demerits

- Shells can sustain only direct stresses and no bending is permissible.
- Shells can take only a negligible amount of shear stresses.
- Shells can be used only as a roof and not suitable as a floor.
- Any damage caused to the shell roof cannot be repaired easily.
- Color washing the shell roof needs special ladder or temporary scaffolding.

4.3   Techniques for Heavy Decks

Erection Methods for Launching of Heavy Deck

- Balanced cantilever erection method.
- Progressive placing method.

- Span by span or Steeping form work method.

- Incremental launching method.

1. Balanced Cantilever Erection Method

In this method of bridge construction is chosen where a bridge has few spans which range from 50 to 250m. Construction begins at each bridge pier. Special formwork is positioned and cast-in-situ pier segment is begun. The complete pier segment is then used as an erection platform and launching base for all subsequent travelling formwork and concrete segment construction. Cast-in-situ segments range between 3mm to 5m in length with formwork moving in tandem with each segment. Segment construction is continued until a joining midpoint is reached where a balanced pair is closed.

Stability of the end cantilever is maintained by using temporary pier supports as the end span is begun. The length of the end spans is equal to between 0.55 and 0.65 times the length of the typical span in the bridge.

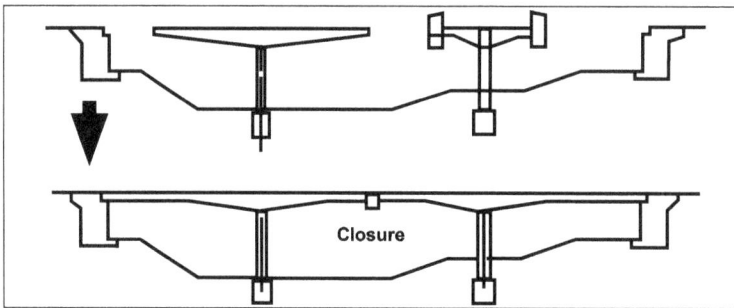

Typical construction sequence.

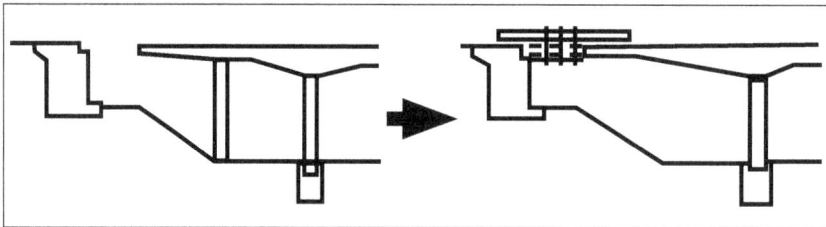

Temporary pier Closure of end cantilever.

2. Progressive Placing Method

The progressive placement method involves starting at one end of the bridge and erecting segments in sequential order. This method of construction is particularly suitable for environmentally sensitive areas or where construction access is limited. It is often called "top-down" construction because the substructure and superstructure can all be erected from the superstructure. The method usually requires the placement of temporary piers at about the middle of each span and is suitable for span lengths of 100 to 300 ft.

Temp. support

3. Span by Span or Steeping form Work Method

In the span by span method, an entire span is assembled, post-tensioned and erected so that it is self-supporting before the next span is erected. The method is appropriate for span lengths up to about 150 ft. Beyond 150 to 180 ft., the method is less cost effective.

In one variation of this method, all the segments are supported by an erection truss before the segments are post-tensioned together. The erection truss may be located either above or below the segments. Once the segments are post-tensioned together and the span is resting on its bearings, the erection truss is moved to the next span. When space permits, the segments may be assembled at ground level, post-tensioned together and the entire span lifted into place.

4. Incremental Launching Method

Incremental launching of bridges can save time, money, space and disruption while easing access and delivering a high quality finish.

The incremental launching method is particularly suited to the construction of continuous post-tensioned multi-span bridges. It involves casting 15-30m long sections of the bridge superstructure in a stationary formwork behind an abutment and pushing a completed section forward with jacks or friction launching system along the bridge axis. The sections are cast contiguously and then stressed together. The superstructure is launched over temporary sliding bearings on the piers. To keep the bending moment low in the superstructure during construction, a launching nose is attached to the front of the bridge deck.

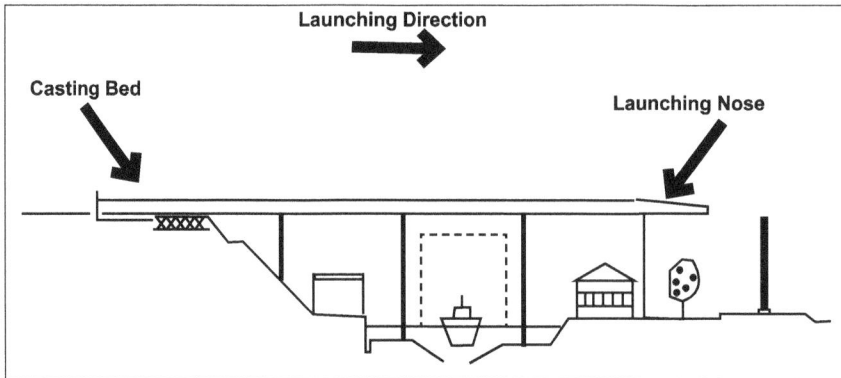

The main advantages for using this construction methodology, rather than other traditional methods, are:

- Minimal disturbance to environmentally sensitive areas.

- Smaller assembly zone required.

- Greater safety during construction which is mainly carried out at ground level.

- Economy of transportation and general reduction in construction elements.

- Higher quality finish and performance derived from easier working conditions and repeatability of tasks.

- Ease of access to restricted or limited sites – such as over rivers, deep valleys and road or train lines, in poor soil conditions or environmentally protected areas.

Although its significant advantages make using this technique a highly attractive option, certain aspects require a high level of expertise both in terms of people and equipment offered by Members of the BBR Network. They have much experience, acquired over many years, in the techniques of incremental launching and a track record for successful delivery of launched structures all over the world.

4.4   In-Situ Prestressing in High Rise Structures

Prestressed concrete is a modified or improved form of reinforced concrete. It takes the full advantage of compressive strength of concrete and at the same time eliminates the weakness of concrete in tension.

Further, the concrete is first subjected to compressive stresses before the external loads are applied. By inducing tensile stresses, external loads are counteracted.

Methods of Prestressing

The two methods commonly used for prestressing are:

- Pre-tensioning method.
- Post-tensioning method.

1. Pretensioning Method

In this method, pretensioning the members is done before the concrete is poured. The cables or tendons are pre-stressed in place in the forms.

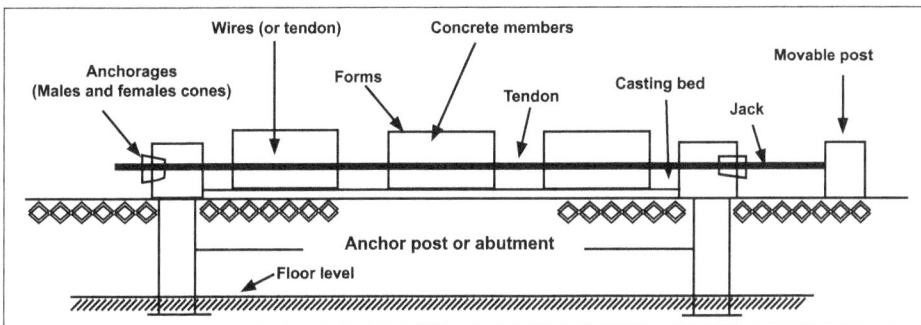

(a) Prestressing bed.

If several members with the same cross-section and identical tendons are to be cast the long-line process is employed. The details of the process are given in Figure (a), (b) and (c).

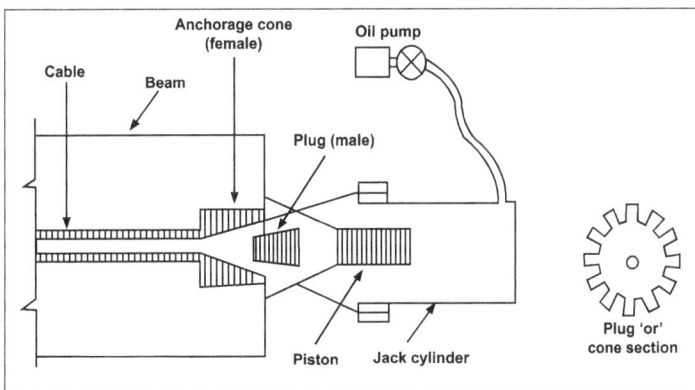

(b) Sectional details at cable or tendon (c) details of tension jack.

The tendons consist of small diameters of high tension steel wires which are first stretched to the desired tension in the forms by hydraulic jacks and then anchored at their ends.

Concrete is now poured in the forms or moulds. When the concrete attains a strength of at least 350 kg/cm$_2$, the tendons are released from their anchorages.

The wires or tendons cannot return to their original position due to the bond with the concrete. This bond induces a compressive force on the member.

The wires at the lower part of the beam, if provided, have subjected to compressive force which is eccentric and results the beam to hog and thereby develops stresses opposite to those that the loads will cause. This method is commonly adopted from pre stressing simply supported slabs, joints and building frames and floors.

2. Post-tensioning Method

In this method, the cables or tendons or wires are placed in position before the concrete is poured. But the prestressing is done only after the concrete attains the desired strength.

In this system, the cables are enclosed in ducts or metal sheets before concreting which prevent them from bonding to concrete. Alternatively, holes are left in the concrete cast at site through which the cables are subsequently threaded.

After the concrete attaining the required strength, the tendons or cables are stretched through ducts or holes by special jacks acting against the end of the site precast member.

After completing the stretching of cables, they are anchored at the ends of concrete member thereby increasing compressive stress in it.

Now, in order to prevent corrosion of cables and improve the overall load-carrying capacity of the member the space around the cables are closed by forcing cement grout under pressure.

Sometimes, the duct is formed of a material such as solid or inflammable rubber or plastic tubing which can be extracted after 12 hours of concreting. The cables are then placed in the duct.

This method is most suitable for in-situ concreting. It is particularly suitable for works which have difficult access. Further, it is preferred for heavy structural members such as bridge girders.

Materials Requirement

In order to obtain the maximum benefit both the concrete and steel should be of high strength. Controlled concrete having a minimum cube strength (of 28 days) of 420 kg/cm$_2$ and 350 kg/cm$_2$ should be used for pre-tensioning and post- tensioning respectively. The steel used in prestressing should be plain, cold-drawn, stress-relieved steel wires or high tensile steel or as drawn steel wires.

4.5   Material Handling

Material handling is the movement, protection, storage and control of materials and products throughout manufacturing, warehousing, distribution, consumption and disposal.

As a process, material handling incorporates a wide range of manual, semi-automated and automated equipment and systems that support logistics and make the supply chain work. Their application helps with:

- Forecasting.

- Resource allocation.

- Production planning.

- Flow and process management.

- Inventory management and control.

- Customer delivery.

- After-sales support and service.

A company's material handling system and processes are put in place to improve customer service, reduce inventory and shorten delivery time and lower overall handling costs in manufacturing, distribution and transportation.

Applications

Material handling systems are used in every industry, including:

Aerospace, Appliance, Automotive, Beverage, Chemicals, Construction, Consumer goods, E-Commerce, Food, Hardware, Hospital, Manufacturing, Materials processing, Paper, Pharmaceutical, Plastics, Retail, Warehousing and distribution.

Principles of Material Handling

When designing a material handling system, it is important to refer to best practices to ensure that all the equipment and processes—including manual, semi-automated and automated—in a facility are working together as a unified system.

By analyzing the goals of the material handling process and aligning them to guidelines, such as the 10 Principles of Material Handling, a properly designed system will improve customer service, reduce inventory and shorten delivery time and lower overall handling costs in manufacturing, distribution and transportation.

These principles include:

Planning

Define the needs, strategic performance objectives and functional specification of the proposed system and supporting technologies at the outset of the design.

The plan should be developed in a team approach, with input from consultants, suppliers and end users, as well as from management, engineering, information systems, finance and operations.

Standardization

All material handling methods, equipment, controls and software should be standardized and able to perform a range of tasks in a variety of operating conditions.

Work

Material handling processes should be simplified by reducing, combining, shortening or eliminating unnecessary movement that will impede productivity. Examples include using gravity to assist in material movement and employing straight-line movement as much as possible.

Ergonomics

Work and working conditions should be adapted to support the abilities of a worker, reduce repetitive and strenuous manual labor and emphasize safety.

Unit Load

Because less effort and work is required to move several individual items together as a single load (as opposed to moving many items one at a time), unit loads such as pallets, containers or totes of items should be used.

Space Utilization

To maximize efficient use of space within a facility, it is important to keep work areas organized and free of clutter, to maximize density in storage areas (without compromising accessibility and flexibility) and to utilize overhead space.

System

Material movement and storage should be coordinated throughout all processes, from receiving, inspection, storage, production, assembly, packaging, unitizing and order selection, to shipping, transportation and the handling of returns.

Environment

Energy use and potential environmental impact should be considered when designing the system, with reusability and recycling processes implemented when possible, as well as safe practices established for handling hazardous materials.

Automation

To improve operational efficiency, responsiveness, consistency and predictability, automated material handling technologies should be deployed when possible and where they make sense to do so.

Life Cycle Cost

For all equipment specified for the system, an analysis of life cycle costs should be conducted. Areas of consideration should include capital investment, installation, setup, programming, training, system testing, operation, maintenance and repair, reuse value and ultimate disposal.

4.6   Erecting Light Weight Components on Tall Structures

Tall buildings such as multi-storeyed buildings, R.C.C.chimneys, grain-storage structures, elevated water-tanks, cooling towers and fireplaces and flues may be called as tall structures.

1. Tall Buildings

Large-scale industrialization has resulted in a great expansion of building programs. Prohibitive land cost in urban areas and demand to meet large population in urban areas, have made way for the construction of tall buildings called as multi-storied buildings. This a stage has reached now that multi-storied construction is essential and inevitable in urban areas.

Buildings with more than five stories are called as multi-storey buildings. Most of the tall buildings in cities have five to twelve stories. But in metros like Kolkata, Delhi, Mumbai and Chennai 20 to 25 stories-buildings have started coming up.

2. Advantages and Disadvantages of Tall Buildings

Advantages

- Economy in use of less land for construction.

- Gives room for large proportion of open space for creating natural environment.

- Enables better day-lighting and greater flow of air.

Disadvantages

- Density of population is high in a small area.

- Prevention of congestion is difficult.

- Excessive and imbalanced load on municipal services like water supply, sewage, electricity, etc.

Erection of Structures

Generally, gantry cranes are used for erection of prefabricated structures from storage depots and pre-assembly areas.

Gantry cranes have a large lifting capacity and adequate stability and are designed for self-assembly. These cranes can be used effectively up to six storeys for larger taller buildings or structures special cranes have to be used.

1. Erection of Multi-storey Frame Buildings

Multi-storey industrial buildings may be built of standardised constructions with column spaces based on some modular grid (say 6 m×9 m or 6 m×6 m). Based on dimensions, multi-storey buildings are erected by means of tower or derrick cranes placed one or both sides of the frame.

When two cranes are put into use, they are to be arranged such that there is no dead area, i.e., areas not served by the cranes. Further, another requirement is that their booms or loads they hoist should not interfere with one another.

This is achieved by positioning the cranes so as to provide a space sufficient for their safe operation. Sometimes the cranes operate in sequence with one lagging behind the other.

When cranes are placed outside the building, erection should be carried out one storey after another. Based on this procedure, the erection of a storey is started only after the completion of all the structures of the preceding storey have been completed.

Further, the erection braces are placed so as to ensure longitudinal stability of the building. When cranes are located within the building cross-section, the vertical break between adjacent cells (units) of the framework should not exceed one tier.

Columns at the ground storey are placed on heads of foundation columns or in foundation pockets. Columns of subsequent storey are mounted using group jigs which are intended for the erection of four or six columns Figure.

A group jig consists of a box-type metal structure with collars for securing columns and a wooden working platform for the erectors. Further, jig carries three collars for each column. The bottom collar is attached to the jig projecting caps of the under lying storey columns.

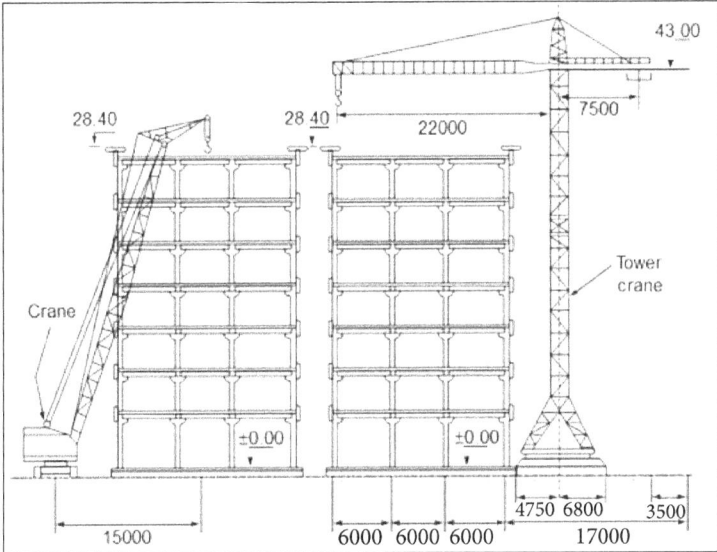

Diagram of locating crane for erecting high-rise building.

The jig is aligned with the center lines with the aid of a special frame. Further, it is secured to erection ears of intermediate floors of bracing. Using screw jacks the jig is levelled.

After the jig has been placed and secured to caps of columns of all underlying storey, all four new columns are installed, secured by adjacent screws. Using a theodolite, the verticality of the columns is checked.

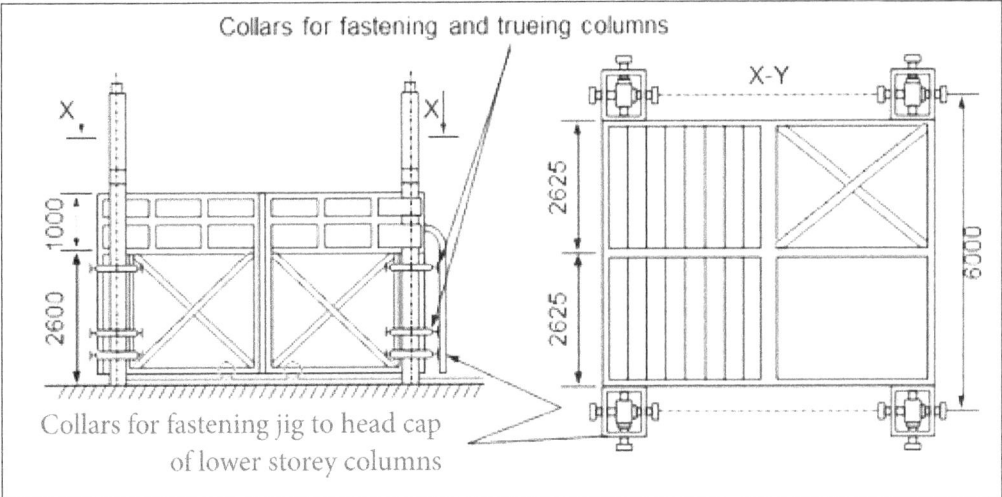

Single-tier for four columns.

Cross-bars of the ground floor are placed then. Once the cross bar has been positioned correctly, its cast-in fittings are tack welded to column brackets.

After securing the cross bar throughout the width of the building, reinforcement projections are welded together. The cast-in fittings of cross-bars and column brackets are permanently welded. The joints are then grouted with concrete.

2. Erection of Tall Steel Structures

In power engineering and industrial construction, use is made of tall steel structures having relatively small load-bearing areas but with both large mass and height.

Power-transmission line towers, radio antenna and TV antenna supports, mobile-communication towers, radio-relay masts, stacks, vertical equipment for metallurgical, chemical, petro-chemical and other industries are the type of tall metallic structures.

Based on the type, mass and size of a structure the erection steps are as follows:

- Lifting or tilting a fully assembled structure to the required position by rotation about a point.

- Lifting a fully assembled structure into the required position by a forcing-out technique.

- Placing a structure into the required position by building it up from components using a climbing crane or an erecting mast.

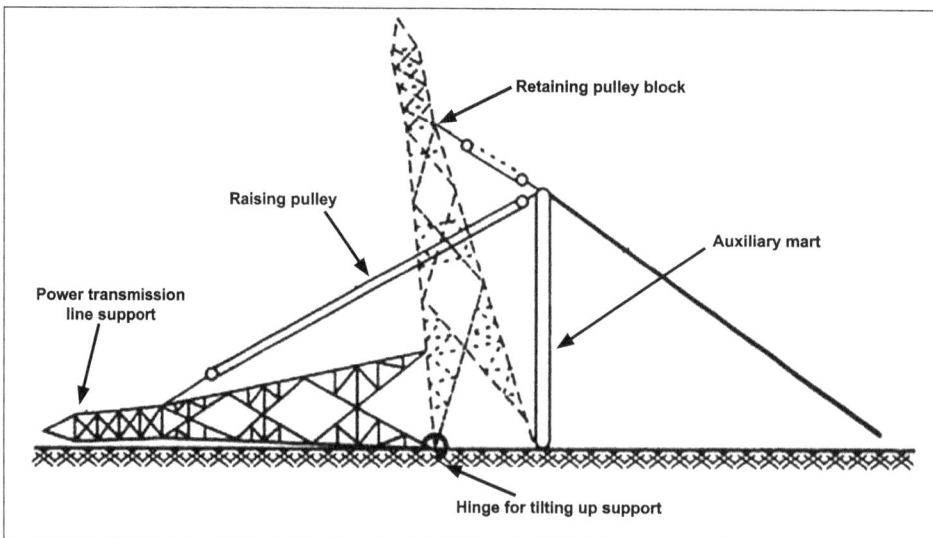

Raising power transmission line support.

The tilting about a supporting pivotal method is used to erect a power transmission line supports. At first the pivotal bottom is hinged on to the same foundation where the lifted structure is to be installed. The end of the boom is attached by a rod to the structure being lifted by a detachable pin on the boom's cap. At the other end, the boom is supported by a lifting compound pulley. The active end of the boom is pulled towards the ground. Thus, the structure is pulled to the required position.

4.7 Support Structure for Heavy Equipment and Conveyors

Support Towers

Many different size towers are available for support purposes. From 2' x 4' towers up to 18' x 20' and beyond are designed and ready for our application. We often support our bucket elevators using our square towers. Most towers are built out of 4" x ¼" square tubing. Stairs and full platforms are available on tower structures as well.

Support towers.

Conveyors

Conveyors are transportation devices which function is adopted the friction between the materials being transported and the base of the conveyor called the belt. These are suitable if the path for the flow of material is fixed. Conveyors may be used for horizontal, vertical (also called as elevators) or inclined conveying materials.

Conveyors may be classified as:

- Belt conveyors.

- Roller conveyors.

- Chain or cable conveyors.

- Pipe line conveyors.

- Elevating conveyors.

1. Belt Conveyors

It is one of the most common form of materials handling system adopted in construction projects. In construction projects, the belt conveyors are used for handling the materials in crushing and screening plants. These are capable of conveying large quantities of materials continuously over long distances with a fast speed. Belts used are made of rubber covered cotton or rayon lay in piles. The capacity of plies to be used is dependent on the strength required for the material to be conveyed.

Belt conveyors consist of a belt running over drums or pulleys provided at the end and are supported at intervals by a series of rollers known as idlers. These idlers in turn are supported on frames. Idlers support the conveyors and reduce the sag of the belt. Otherwise, there will be the possibility of spilling of the materials being conveyed.

2. Roller Conveyors

These types of conveyors are with rollers which are supported on frames over which materials are allowed to move. They may be driven using power or by gravity. Different types of roller conveyors are available which may be used to move materials in horizontally or vertically downwards. These are used to transport various types of materials such as boxes, tiles, etc.

3. Cable Conveyors

These types of conveyors are moved by chains or cables in horizontal direction. They are fixed just above or on the floors. These types of conveyors are used for moving heavy boxes of materials and grates of big boilers.

4. Pipeline Conveyors

This is mostly used for moving dry materials like sand, cement, chemical powder, etc. This is operated by gravity or by air pressure or by some mechanical means.

5. Elevating Conveyors

These are also known as bucket elevating conveyors or bucket elevators. They carry materials vertical or near vertical positions.

There are two types, viz.:

- Chain bucket elevator in which buckets are attached to one or two chains which move on two end wheels.

- Belt bucket elevators in which buckets are attached to the belt which is moving on pulleys provided at two ends.

4.8 Erection of Articulated Structures, Braced Domes and Space Decks

Articulated Structure

Articulated structures means the separation of a structure into two or more elements and join the entire structural elements such that it functions as a single monolithic structure. The structural elements are prefabricated and are assembled and erected.

Design and Manufacturing

While designing prefabricated buildings and installations high technological effectiveness should be taken into account. The design of structural parts, utilization of structural parts and their joints should be installed with minimum use of materials and manpower for manufacture and erection.

In fully prefabricated construction, it is the practice to use larger elements while simultaneously reducing their relative mass. This is achieved by using more efficient design, light weight concrete, synthetic heat insulation and other efficient materials.

At present, prefabricated concrete factories which not only manufacture structural components but also assemble buildings from prefabricated blocks and perform the whole complex of construction work. As prefabricated elements grow larger and fuller prefabrication makes possible speedier construction.

The erection should involve minimum consumption of labour, time and other means. Effectiveness in erection depends or efficient pre-assembly of structures, relatively equal weights of erection units, high degree of prefabrication and accuracy of manufacture and simplicity of but joints and provision of fastening devices.

Delivery and Storage of Prefabricated Structure

Structures are delivered to erection zones by most effective mechanized procedures and allowing for haulage distance, availability of approach roads and conditions of in-site roads.

Depending on the character and the application of, structures, they are transported from manufacturing work to construction sites and unloaded at a pre - assembly area. Pre- assembly area is a site storage area or a zone identified in an erecting area.

Prefabricated structures are generally transported by trucks with two- axle trailers, tractor trucks with semi-trailers, trailers and panel transporters.

Delivered units to the construction site should correspond to assembly lists which specify the name, the type and the number of prefabricated elements intended for placement in a specified area of the building.

Pre-assembly of Prefabricated Metal Structures

Haulage of large size building or other structures may be cumbersome because of their size, mass and difficulty in transportation. These structures are transported from the manufacturing works in the form of transportable components which are assembled into erection units at the construction site.

Metal structures are generally pre-assembled on special metal racks up to 80 cm high built of uprights carrying rails, I- beams and channels. Sheet structures of round configuration are assembled into circular sections or blocks of 2 or 3 circular sections. If too large for convenient handling, the structures are preassembled on an area covering by erection cranes.

Space Decks and Bridge Decks

Bridges are designed for carrying, moving loads and personnel. They need a flat surface for movement. This flat surface is called the deck surface. The deck may consist of a slab, a beam and a slab, a grillage, a box-girder, multi concrete beam, etc. The simplest form of a concrete bridge deck is a solid slab which could be reinforced or prestressed.

Bridge deck slabs are normally supported at two edges and the remaining two are free. Space decks are those which are used for launching of space vehicles and as decks on off shore structures. Space decks are to be designed for dynamic stresses in addition to static stresses.

5

Construction Equipment

5.1   Selection of Equipment for Earth Work

The major factors which govern the selection of equipment for earth work are:

- Physical job condition.
- Type of soil.
- Specification of machine.
- Condition of the machine.
- Method of operation.

Equipment Cost

Procurement Cost of Investment

The procurement cost consists of the following factors:

- Cost price.
- Interest on money invested.
- Taxes.
- Insurance.

Operating Cost

The operating cost of equipment is based on the following factors:

- Cost of investment.
- Depreciation cost.
- Cost of major repair.
- Cost of fuel and lubricants.
- Cost of labour.
- Servicing and field repairs.
- Overheads.

Cost of investment: The amount invested for the purchase of equipment is to be compared with an alternate investment made in some industry or deposited in bank.

Depreciation cost: Because of wear and tear the cost of equipment reduces which is called as depreciation cost. Depreciation is also a method of assessing the smooth running of the industry. An amount of the earnings has to be set aside so that the plant or industry required a complete change at that time the accumulated amount can be used.

Cost of major repair: Major repairs are those which are incurred when the equipment is taken to a workshop. The major repair involves replacement of machine parts, overhauling or servicing. This expenditure is added to the capital cost so that the expenditure is distributed over the years and included in the operating cost of the machine.

Cost of fuel and lubricant: The possible consumption of fuel for the full load condition and constant speed under favorable conditions are considered while calculating the fuel consumption. As the equipment will not be working at this ideal condition, reduction factor of 60% is considered.

Cost of Labour

The cost of labour includes:

- The salaries of working crew (operator, mechanic, helper, etc.).
- Salary of supervisory staff.
- Salary on account of leave salary.

Servicing and Field Repair

All expenses involved in the maintenance normally met during the operation of equipment except major repairs are included under this. This includes periodical service of the equipment.

Overhead

On a large project, the whole establishment is work charged. Any unforeseen expenses which don't work charged or nonproductive are distributed to the various work producing elements for the purpose of cost control. No machine will exactly satisfy the 100% need of the work. As a matter of fact, a large number of factors are contributing to the best performance of a machine.

A lack of knowledge of some or all of these factors may affect the efficiency of operation of particular equipment. While designing earth-moving equipment, a designer has to make a compromise to suit his design to satisfy the major conditions of work. Thus, it is not possible to get the same efficiency under different set of conditions.

5.2 Earth Moving Operations

There is a wide range of excavating equipment available. In selection a much greater care and thought needed to find a suitable machine for a particular job. The best performance of a machine depends on a large number of factors.

It should be noted that no machine is designed to suit a particular set of conditions which may be demanded at a construction project. Thus, a best designer would design the machine such that the machine fits to most construction projects.

Excavators

These are digging machines. These machines consist of the following components:

- An undercarriage: This gives mobility to the excavator. This may be mounted with crawler track or wheel.

- A superstructure with operator's cabin: This could traverse through 360° or fitted on a rigid frame.

- Hydraulically articulated booms and dipper arms with bucket.

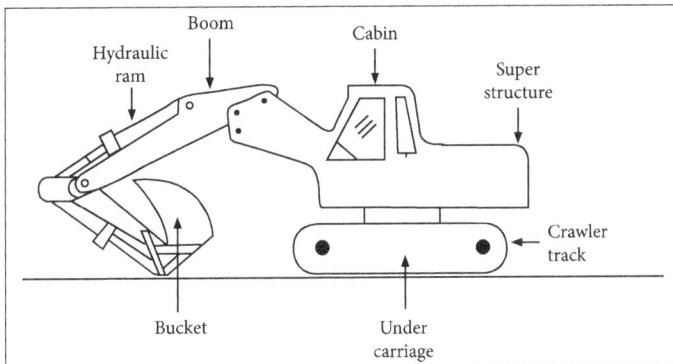

Excavator.

1. Excavator-loaders

These loaders are also known as back-hoe loaders. Primarily, they are digging and loading machines. It consists of:

- A tractor with two-or-four-wheel drive.

- A rear mounted digging arm and hydraulically operated excavator bucket.

- A front mounted loader arm assembly with loader bucket.

Excavator end is used for side tipping, excavating for building foundations, irrigation canals, excavation for pipelines, etc. The loader end is used for loading and re-handling of materials such as excavated soil, aggregates, coal, etc. Further, it can be used by adopting hydraulically operated attachments for road breakers, pumps, pole planters, winches, crane hook and boom, etc.

Excavator-loader.

2. Bucket-wheel Excavator

Bucket-wheel excavators are provided by digging wheel having buckets. These buckets can be vertically adjusted by means of hydraulic rams. The wheel is mounted as a boom having a conveyor belt on which this wheel discharges. This wheel can swing a complete 360° circle. A second conveyor which is co-linear with the first receives the material from the first conveyor.

The second conveyor can be sleeved through 180°. Further, it can be raised and lowered hydraulically to accommodate the receiving of handling units at varying heights. These excavators are available with the capacities of 200 to 11000 cum. These wheel excavators are used in mining work.

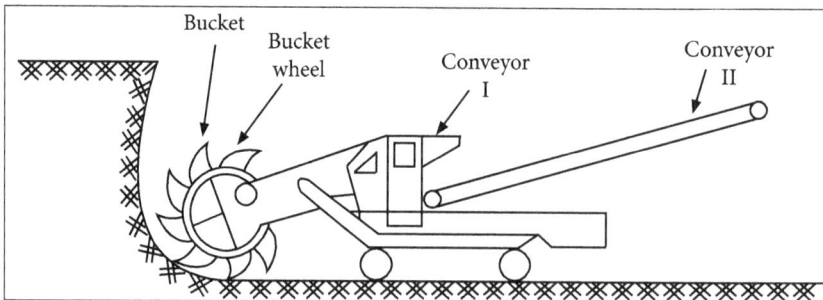

Bucket wheel excavator.

5.3   Types of Earthwork Equipment

In recent years, a great deal of improvement has been made in earth moving equipment.

Earth moving equipment may be broadly classified into:

- Excavating machines.

- ◦ Tractor and tractor units.
- ◦ Scrapers or pans.
- Grading and compacting machines.

5.3.1 Tractors

A tractor is a multi-purpose machine. This comes in varied types as a light model to heavy model. The light model is used for agricultural or small haulage purposes. Heavy model equipped with several special rigs is used for earth moving work. This is an important piece of equipment which is indispensable on all important projects.

Two principal applications of tractors are:

- Clearing and excavating machinery.
- Hauling and conveying machinery.

There are two types, viz., wheeled tractors and crawler tractors. The wheeled type is used for light and speedy jobs. As regards to its applicability, it falls between the crawler tractor and the truck. The crawler tractors are rugged machines which are used for heavy duty work. It is used particularly when there is a demand for more tractor power and speed of movement.

Now a days wheeled units have been made to work on the jobs which are intended for the crawler type. Wheeled tractors are now available for all practical earth moving jobs, including ripping and dozing.

Important Qualities of Wheeled Tractors

- These can travel faster than crawler tractors and can be used where long distance to be moved with high speed.
- These have wheel steering control as in automobile vehicles and are easily measured.
- These can be self-driven over long distances.
- These tractors need less skill of operation and less maintenance cost.
- These are comparatively less in cost.

Important Qualities of Crawler Tractors

- These are more compact and powerful which can handle heavy jobs of hauling and digging.

- Thus, when used for long distances are done with trailers.

- These are not liable to slip over very smooth or loose footings when increased power is applied to the wheels.

- These fitted with special shoes while it is moved on pavements or on tarred roads.

5.3.2 Motor Graders

A grader is primarily a device for leveling or finishing earthwork. Sometimes, it is also used for mixing gravel, making windrows and trimming slopes.

There are two types of graders, viz., towed and motorized.

The towed type is of small size with a tractor. The controls are in general manual and sometimes a small petrol engine is fitted on the framework of the grader to operate the controls. It is operated by a separate operator.

The power to operate the controls is supplied through a quill shaft passing through hollow clutch and transmission shafts. Irrespective of the movement of the machine, the quill shaft is made to rotate whenever the engine is running.

Attachments to the grader include the blade also called as moldboard, the scarifier, the bulldozer or the snowplow, the elevator and the roller. The versatility of the machine is increased by the addition of these tools.

The blade is thick and can swivel through 180°. The scarifier is the tool for loosening hard soil and may be mounted in the front or in the rear of the machine.

The bulldozer and the snow-plow a V-shaped blade are front-end attachments fitted to the grader for pushing loads or clearing snow.

The elevator attachment enables the grader to pick up the material cut by the blade and drop it over an inclined belt conveyor which transmits it into the carrier unit or discharges it aside the window. In order to compact or smoothen the graded surface, roller may be attached to the rear of a grader.

5.3.3 Scrapers and Front End Waders

Scrapers are the devices to scrap the ground to load the material, to transport to the required distance, to dump at the intended site, to spread the dumped material over the required area, to attain the desired thickness and to return back to do the next cycle.

In simple terms, scrapers are designed to dig, load, haul, dump and spread. As a scraper does a multiple works it is also called as carry all.

1. Parts of Scraper

A scraper consists of the following parts:

Bowl

It is a pan which is to hold the scraped material. It is hinged at the rear corners to the rear axle inside the wheels. It can tilt down for digging or ejecting. The size of a scraper is specified by the size of the bowl. At the bottom of the bowl a cutting edge is attached. In order to make a shallow cut, the cutting edge is lowered into the material or dirt.

Apron

It is a wall located in front of the bowl to open or close in order to regulate the flow of the material in and out of the bowl. Further, it is capable to open or close during the carrying position also.

Ejector

It is also called as a tail gate which is the rear of the pan. It is provided with forward and backward movements inside the bowl. During loading it remains at its rear wall and moves in the forward direction to help in the ejection of the load during dumping.

Hydraulic System

All these operations are controlled by hydraulic cylinders.

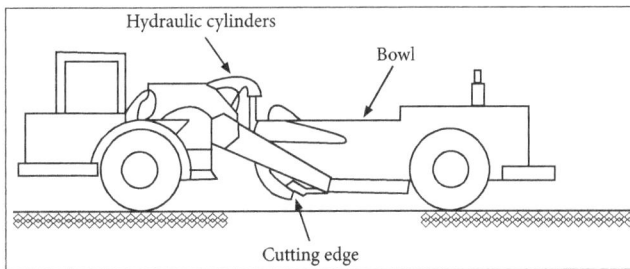

Scraper.

2. Operation of a Scraper

The operations of a conventional scraper are given below:

Digging or Loading

Keeping the ejector at the rear and the apron raised (approximately to 40 cm) the operator moves to the cut. The bowl is then lowered to the required depth of cut, the engine speed is increased and moved forward keeping the optimum depth of cut. After filling the bowl, the apron is closed and the bowl is then raised.

Transporting

The bowl is transported in the raised position so as to provide sufficient clearance from the ground. In order to prevent the loss of the collected material, the apron is fully closed.

Unloading

The process of unloading by the scraper is called as dumping and spreading operation. In this operation, the bowl is positioned so as to spread the material. So, as to have an even spreading, a partial opening of the apron in the stage of unloading will be helpful. In the case of wet and sticky material, the apron should be raised and lowered repeatedly.

This helps the material behind it is get loosened and material drops out of the bowl by moving the ejector forward, the remaining material is pushed out at a uniform rate. After the dumping is complete, the tail gate is fully retarded, the apron is dropped and the bowl is raised to the transporting position. Using the cutting edge final finishing is done.

3. Types of Scrapers

- Towed Type.

- Self-Propelled.

- Self-Loading.

- Tandem Scraper.

- Twin Engine Scraper.

- Multi-engine, Multi-Bowl Scraper.

5.3.4 Earth Movers

Bulldozers

It is important equipment on a construction project. It is basically a scraping and pushing unit. However, it is multipurpose equipment which can be used for different purposes with some modifications. Accordingly, they are called as angle dozer, tilt dozer, tree dozer and push dozer. Up to 100 m distance these can be used to haul.

Angle dozer pushes its load at an angle (nearly 25°) to the direction of travel. This is helpful:

- When the material has to be pushed down the slope on hill work,

- Where a long windrow has to be made during the travel of the dozer.

Tilt dozer is used to start excavating a ditch or a trench or for excavation in hard ground. For this, the blade is required to be tilted by raising one end up to 25 cm above the other. Push dozer is used to push the scraper unit after digging and also during loading operation using the pusher plate. Tree dozer is used to uproot and remove trees.

5.4   Equipment for Foundation and Pile Driving

Pile driving equipment comprise of the following:

- Driving rig.

- Guiding leaders.

- Pile hammer with accessories.

- Additional aids for pre boring and jetting.

- Boiler for steam raising or air compressor.

1. Driving Rigs

Driving rig provides basic operations of lifting the pile, holding the pile in position, hammering it into the ground or of pulling it out of the ground and guiding the pile in the desired direction of movement. The rig supports the boom and winch mechanism, driving hammer, the guiding leaders and a platform for mounting of auxiliary equipment such as a jet pump, drilling auger, steam boiler or air compressor.

2. Guiding Leaders

The leaders guide the pile and the hammer during an operation which extends to the entire height of the rig. In case of piles to be driven below, the level of the rig into excavations, trenches or water telescopic or extensible leaders can be used. The leader should enable the hammer to deliver blows axially to the pile.

During the process of driving, the driving rig should be strong and stable. In case a boom is used, adequate space should be available between the pile top and the point of the hammer to work. There are two types of rigs, viz., skid–mounted and crane-mounted.

The skid–mounted rigs are provided with rail wheels or with long steel rollers for movement.

The crane mounted rigs are mounted with a crawler or a truck chassis with a swinging deck. In the case of floating pile driving, both the rigs can be mounted on a barge.

3. Driving Hammers

Pile driving hammers impart energy required to drive the pile into the soil. The routinely used pile hammers work by hitting the pile on its head.

The vibratory and sonic types are two new types of hammers.

Hammers are classified as:

- Drop hammers.
- Single acting hammers.
- Double acting hammers.
- Differential acting hammers.
- Diesel hammers.
- Hydraulic hammers.
- Vibratory hammers.
- Sonic hammers.

Drop Hammers

This is the simplest form of hammer which does not use any external source of power. The only mechanism needed is to lift the hammer through a cable. Although the process is slower, it is more efficient as it uses only the gravity. The drop hammer is basically a block of suitably shaped cast-iron with its center of gravity centered near the base in order to facilitate smoothness of fall.

Single Acting Hammer

The functioning of single acting hammer differs from drop hammer only in the manner of lifting of the ram after each blow. A conventional single acting hammer employs a piston connected to a ram at its bottom end and moving inside a cylinder. The hammer may be of an open type or closed type.

Stream power or compressed air is used in the single acting hammer without any adjustment or alterations and the pressure remains unchanged. This pressure ranges from 5.6 to 10.5 kg/sq.cm and used depending on the size of the hammer and its weight. The operation of the single-acting hammer costs less compared to a double-acting hammer but its speed is slower.

Double Acting Hammer

In this type of hammer, the motive fluid acts on both sides of the piston. This assists the force of gravity in striking the blow by raising the ram to the top of the driving stroke and then driving it down in its down stroke.

An equivalent energy of the single acting hammer can be developed by dropping the ram through a shorter distance and thereby reduces the energy loss. Stream or compressed air is used as the motive fluid.

Depending on the hammer weight and the weight of the ram, these hammers are available in light, medium and heavy duty models.

Two types of hammers are available, viz., open end and close end. The close-end model is used for underwater driving without followers. This hammer is capable of fast driving in light or medium heavy piles in moderately hard soils.

Its advantage is the speed and the disadvantage is that it should be supplied with a stream or air at the specified pressure. It is especially suitable to drive steel piles.

Differential Acting Hammer

Advantages of single acting and double acting hammers are combined with differential acting hammer. But, it suffers with their disadvantages. That is, it uses a heavy ram weight, as followed in single acting hammer and a high speed of operation, as followed in double acting hammer. Thus, it derives the advantage of use under a wide range of pile driving conditions.

It uses 25 to 30 % of less steam compared to a single-acting hammer. This hammer is available in open and closed models and could be used, using either steam or compressed air. Lubrication system is similar to that used in single acting hammer. The principle advantage of the differential acting hammer over the double acting hammer is its larger ram weight. Due to the wide range of ram weight and operating speed, it suits any type of pile and field conditions.

Diesel Hammer

The working of this hammer is different from other types. It does not depend on the motive fluid, but has a self-contained power source. This is more efficient as the hammer is designed to deliver driving energy to the pile in three forms. This is not efficient in hard driving or in extremely cold weather.

Hydraulic Hammer

In this type of hammers, the differential principle of hydraulic fluid is utilized in place of steam or compressed air. It uses a high pressure, which comprises of oil pump, piston, cylinder, fluid lines and fittings, etc. This arrangement reduces the volumetric requirements of the unit and results in a compact hammer.

Because of the use of hydraulic pressure, more driving force and ram speed are available with high economics is achieved. It has the advantages of lower set up and moving costs, saving in time due to absence of steam boiler or air compressor, lower fuel costs, increased speed and easier handling due to the compactness of size. Further, the hammer performance is unaffected by the reactions of the pile being driven.

Vibrating Hammer

With the use of conventional driving hammers blows are struck on the pile head with certain frequency and due to this the pile moves down. Friction develops on the surface of the pile before the next blow. This could be overcome if the pile is driven continuously.

The vibratory hammer takes care of this and allows the pile to move continuously, thereby eliminating surface frictional resistance. In this process, the pile sinks fast into the soil due to its own weight and the weight of the driver assembly.

The hammer comprises of a vibratory unit which produces vibrations to oscillate the pile along its vertical axis and a clamping device which transmits these vibrations to the pile. The driving powers for the vibrators are given by internal combustion engines or with electric motors. Vibrations are generated by movement of counter rotating shafts to which eccentric weights are fixed.

Sonic Hammer

It works on the principle of vibration with some modification. In this, the frequency of vibration used in driving a pile is its resonant frequency. When a device is powered by sonic waves it expands and contracts in proportion to the frequency of vibration.

Thus, when the pile is energized to its resonant frequency, it alternatively expands and contracts a small amount depending on the speed of the sound in the material of which the pile is made.

Expansion of pile at the tip displaces the adjacent soil. Further, the contraction of the pile reduces the friction and because of this pile sinks automatically. Thus, the sonic effect is double fold and because of this the pile is inserted at a fast rate.

Frictional resistance is developed at the soil - pile interface. Frictional resistance can be improved by:

- Increasing the density of the soil which is more possible in granular soil.
- Increasing the roughness of the pile surface.
- Cast in situ piles in all type of soils.
- Driven piles in granular soils.
- Increasing the surface area of the pile.

5.5 Equipment for Compaction, Batching, Mixing and Concreting

Types of Rollers and Compactors

Compaction equipment mainly consists of static rollers and other vibratory compactors. These rollers and compactors may be grouped into the following major classes:

1. Static Smooth Wheeled Rollers

These rollers are used with or without ballast and may be provided with three wheels or two wheels of equal width called tandom type. These rollers are generally used for most of the work. But, these rollers are not effective on uniformly graded sand, gravel or silt and on cohesive soil with high moisture content due to poor traction.

These static rollers, also called as dead weight rollers, are diesel powered. These rollers rely on the weight only to compact the materials by passing over them. Units of 8 to 10 tonnes can impact a pressure of 20 to 40 kg per linear cm are generally in use. Rollers with weight up to 1 ton are used for light work.

2. Sheeps Foot or Pad Foot Rollers

These rollers are suitable for cohesive soils. These may be self-driven or tractor driven and especially useful when the water content is on the higher side. The mass of the drum can be varied by adding ballast. For effective rolling, the lift thickness should be small and the contact pressures under the projection are very high. These rollers are specially recommended for water-retaining earth works.

3. Pneumatic Tyred Rollers

In pneumatic tyred rollers, wheels are placed close together on two axles and placed such that the rear set of wheels overlap the lines of the front set to ensure complete coverage of the soil surface. In order to avoid the lateral displacement of soil, wide tyres with flat treads are provided. The compaction produced by these type is better than that of the smooth wheel rollers.

Static Compaction Equipment

These rollers can be used on any type of compaction, in general. Light weight rollers of 3 tons for foot path construction whereas for road construction heavy rollers are used. Static rollers with weight of 8 to 10 tons are used for works ranging from earth work and sub-bases to bituminous road surfacing materials. Tandom roller is preferred as a finishing roller on wearing course.

Static compaction rollers and compactors may be classified in to the following groups:

- Towed static smooth compactors.
- Static sheeps foot or pad foot compactors.
- Static three wheel self-propelled compactors.
- Static tandom compactors.

1. Towed Static Smooth Compactors

These are the old type of compactors. These are the first type of rolling compaction equipments used. These were pulled by men or horses. During the rein of Romans, these were used to smoothen roads and paths.

2. Static Sheeps foot or Pad Foot Compactors

On the surface of the roller projection resembling that of sheep's foot, i.e., club-shaped and tapered are provided. These feet are also called as lugs which are of different shapes. When earth fills are compacted, the feet penetrate deep into the loose material during the first pass and compact the soil from the bottom up. Sheeps foot rollers provide a kneading effect and break the earth lumps and thereby reduce air voids in a cohesive and in other fine grained soils.

3. Static Three Wheel Self-propelled Compactors

The compactors have three rolls, a small split steering roll in the front and two large drive rolls mounted on rear axles at both the ends. These rollers weigh 8 to 10 tons. This is the most used type applications for a variety of works, viz., earth work compaction, sub - base or base course rolling and for bituminous road rolling.

4. Static Tandom Compactors

Tandom compactors have two equal sized rollers and one centered in line tandom. As these rollers have a smooth surface, they are suitable for compacting bituminous layers. Improvements have been made on these types of compactors as tandom vibratory compactors. Large size tandom vibratory compactors are generally preferred now a days as they can be used either as static compactor or as a vibratory compactor as per the requirement.

Vibratory Compaction Equipment

Vibratory compactors can be categorized into the following groups:

- Tandom vibratory compactors.
- Towed vibratory compactors.

- Towed sheeps foot and tamping–foot vibratory compactors.

- Self–propelled vibratory compactors.

- Hand-guided vibratory compactors.

1. Tandom Vibratory Compactors

Two types are available, viz., single–drum vibrating or double-drum vibrating. In the compactors with double-drum vibrating system, two tandom wheels are provided with separate controlled vibrators in the front and rear rolls. Comparing single and double drum vibratory compactors, the output of double drum vibratory compactor is to be 80% more than the single–drum vibratory compactor. The double-drum vibratory type has an option to operate the single drum or the double-drum.

2. Towed Vibratory Compactors

This type of compactor is especially used for compacting cohesive soils, fine and coarse grained mixed soil and rock materials. The heavy type towed vibratory compactor is used in earth dam and embankment constructions. Because of the large amplitude, it shows more impact motion and therefore preferred for compacting cohesive soils, fine grained soils and mixed fine or coarse grained soils.

3. Towed Sheeps foot or Tamping–foot Vibratory Compactors

This type of compactor is useful in highly cohesive soils and soft rocks. The kneading and crushing effect of the feet improves the compaction performance. The compactors are provided with sheeps foot or tamping–foot vibratory equipment. Tamping-foot are larger than the sheeps foot and hence has a more contact area.

4. Self–propelled Vibratory Compactors

These compactors are available with a weight of 8 to 10 tons dead weight. In one type, large vibratory steel roll in the front and two rubber tyres at the rear. The rubber tyres may be smooth or with treads. The smooth one can be used for bituminous work. In the other type vibratory steel roll is in the front and two static steel rolls at the rear of the multi-purpose work.

5. Hand-guided Vibratory Compactors

Compactors of this type have duplex or double vibratory compactors with dual roll vibration and dual roll drive. These vibrators are small in size and thereby enable cross country mobility and excellent grade ability. They have a provision for dual roll drive. These are especially used for compacting trenches, slopes, parking lots, small repair jobs and sports grounds, etc.

Rubber-tyred Compaction Equipment

Pneumatic-tyred rollers are provided with the tyre-configuration of odd numbers, viz., five, seven, nine, etc. with a certain degree of overlap. Even numbers of tyres are provided on the rear axle and odd number on the front axle. Rubber-tyred or Pneumatic-tyred compactors have been used for earth, a works for cohesive and non-cohesive soils. These are generally self-propelled and are built over a wide range of weights.

These types of compactors are very efficient and produce more uniform compacted surface than steel rollers. These are used for stabilised soils in airfield, embankment and load construction applications. These types of rollers have the following advantages compared to steel rollers:

- Surface of the layer is not bridged, but uniform.

- Bituminous layers, compacted by rubber-tyre rollers show better sealed to keep dirt and moisture.

- Post-compaction by traffic is negligible when compacted by rubber-tyred rollers.

Deep Compaction Techniques

Rollers and compactors are useful for compacting materials on the surface. But field conditions may demand compacting at a deeper depth. Some of the vibration methods are Vibro-compaction, Vibro-displacement compaction. Another method which of recent origin is called heavy damping which is discussed below.

The most basic and simplest way of compacting loose soil is by repeated dropping of a weight on the ground. The method is also known as deep dynamic compaction or deep dynamic consolidation. This method consists of allowing a very heavy weight (up to 400 kN) to fall freely on the ground surface from a height of 15 to 40 m. This leaves an impression on the ground.

The tamping is then repeated either at the same location or over other parts of the area to be stabilized. In the case of non-cohesive soils, the impact energy causes liquefaction, followed by settlement as water drain. Fissures formed around the impact points, sometimes facilitate drainage in some soils. This method has been successfully used to treat various types of soils and fill deposits up to 20 m thick. This method can be used on densifying soils both above and below the water table.

Concreting Equipment

Concreting generally involves: batching and mixing, handling and transportation, placing, finishing and curing.

A concrete plant is provided with arrangements for:

- Receiving all ingredients for making concrete, viz., aggregates, sand and cement and water.

- Weighing each ingredient of concrete for each batch of the mix.

- Mixing these ingredients thoroughly to form a concrete of required consistency.

Concrete Batching and Mixing Plant

Batching comprises of proportioning of ingredients of concrete, viz., aggregate, sand, cement and water, separately for each batch. In the construction field, batching and mixing plants consists of the following:

1. Aggregate Feeders

Aggregate feed bins are made available for each size of aggregate and sand and mix based on volume. In important jobs, weighing system is adopted.

Aggregate feed bins are loaded by the following methods:

- By shovels directly into the bins.

- By Lorries tipping directly into the bins.

- By means of boom scrapers from the aggregates stored in bulk heads on the ground.

2. Cement Silos

Cement is stored in silos. Cement is filled to the silos by the cement carrier by pumping. For each batch mix, required cement is got after proper weighing. This weighed cement is carried to the mixing unit through the enclosed conveyor belts.

3. Water

Water is an important ingredient which decides the quality of concrete. A proper quantity of water based on the water cement ratio has to be added. Because of this the water is to be measured correctly.

The measured water from the measuring tank is delivered to the mixer when already aggregate, sand and cement are charged into the drum of the mixer. Thus, after each charge the measuring water tank is filled up from the water tank to the required quantity.

The water from the measuring tank is supplied to the mixer through the adjustable spray bar so as to attain homogeneous mix within a shortest time.

4. Mixing Unit

Mixing unit consists of two steel parallel shafts provided with adjustable paddle mounted on external supports provided with bearings.

The mix from the mixer discharges into a hopper for delivering into the concrete dumpers directly or into the bin where the mix is stored temporarily.

The mix is then taken to the construction site. The quality of concrete produced following the above procedure is very high quality.

5.6   Equipment for Material Handling and Erection of Structures

Material handling is an important function in a construction work. There is a need of equipment to handle heavy loads with fast speed, reliability, safety and economy. The main objective of the efficient material handling is to save costs.

Material handling devices, are expected to satisfy one or more of the following functions:

- Construction material is to be moved and positioned.

- Lifting of a load and placing it at a particular place.

- Loading of materials into transportation equipment.

- Unloading of materials from transportation equipment.

Type of Handling Devices

1. Vertical Motion Devices (Lifting and Lowering Devices)

- Block and tackle.

- Winches.

- Hoists.

- Elevators.

2. Horizontal Motion Devices (Transportation Devices)

- Wheel barrows and hand trucks.

- Narrow-gauge mine rail road.

- Tractors and trailors.

- Skids.
- Pipe line.

3. Combination Devices (Lifting/Lowering/Transportation)

- Spiral chute.
- Lift truck.
- Fork lifts truck.
- Cranes.
- Conveyors.

4. Aerial Transport

- Cable ways.
- Rope ways.

Vertical Movement Devices

1. Block and Tackle

This is a vertical distance movement device which is the oldest and the simplest device. It depends on mechanical power and gives only mechanical advantage. It is most inexpensive device but waste of manpower.

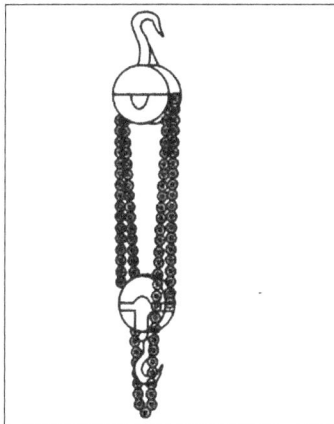

Block and tackle.

2. Winch

By winding the rope of cable on the drum vertical movement is attained. Manpower or other power can be used to wind and a greater mechanical advantage than that of block and tackle. It is often used to load heavy equipment into ships construction equipment, etc.

Horizontal Movement Devices

1. Narrow-gauge Rail Road

As it is very expensive, this is adopted only in very large projects. It is used in industries like blast furnace, copper refineries and steel-rolling operations.

2. Tractors and Trailers

These are the commonest modes of horizontal transportation. Trailers can be lift-loaded and can be towed by tractors. Different types of trailers can be picked up by tractors. This is one of the commonly used methods of handling materials from one place to another. This is less costly compared to Combination Devices.

Lift Track

These are similar to roller skids but provide provision for a large platform to lift and place the material and move them horizontally through power to another location.

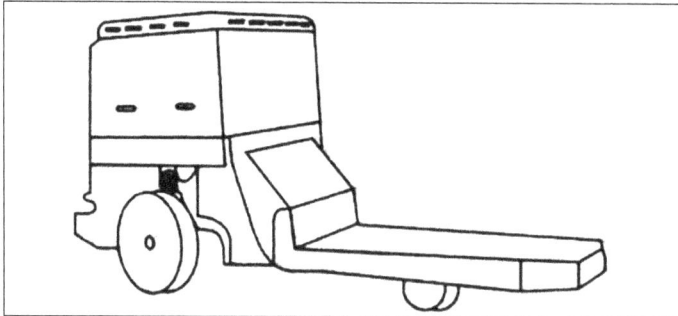

Lift truck.

Fork Lift Truck

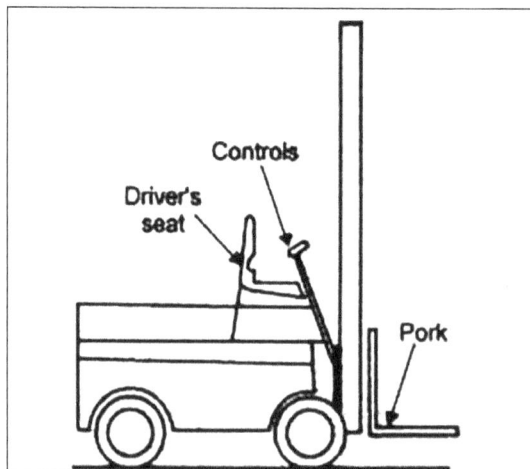

Power-operated fork truck.

It is provided with fork which receives the load at ground level and elevates it hydraulically to the desired height. There is no need for manual lifting. Self- loading or un-loading can be carried out by providing a fork at the front end of the truck. Fork lift trucks are used in construction industry.

Conveyors

These are the material transportation devices used when the parts of the flow of material are fixed. Because of this desired fixity lifting and lowering of materials are done automatically. Conveyors require no stopping or starting but the operation is continuous. The transportation is affected by friction between materials being transported and the belt or roller.

Aerial Transport

1. Cable Ways

In cable ways, during the process of travel the load can be raised or lowered at any point. This is very useful in excavation work of dam, quarries, construction, etc. Here, the loads are hoisted and moved horizontally. These are also used for conveying concrete and placing if the quantity to be concreted is very large.

2. Rope Ways

These are used for long distance movement of materials. Rope ways have two end towers with or without intermediate towers. There are two types. In the first type, endless rope runs over pulleys or horizontal sheaves at the two end towers and is supported along its length on a series of pulleys mounted on intermediate towers. In the second type, separate handling and support ropes are provided. It is costly, but has large carrying capacities.

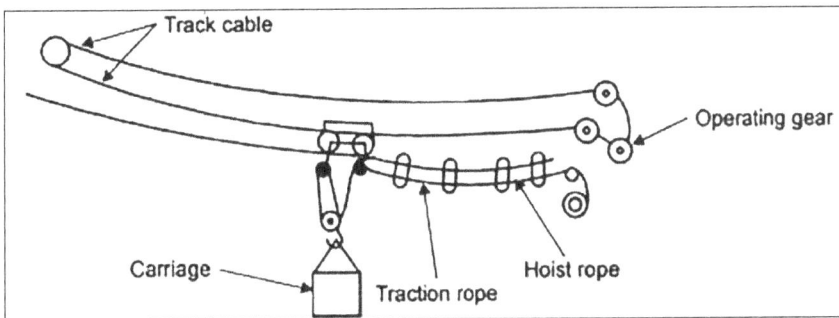

Rope ways.

Erection of Structural Steel Girders

1. General

Before taking possession and erecting the girders, the Contractor shall verify that the lengths of the girders, the layout of the substructure units, the elevations of the bearings seats and the location of the anchor bolts are in accordance with the Drawings.

All discrepancies discovered by the Contractor shall be brought immediately to the attention of the engineer. It is essential that the girders be erected with utmost attention being given to girder positioning, alignment and elevation.

The Contractor shall adjust girder position, bearing location and bearing elevation in order to achieve as closely as possible the lines and grades shown on the Drawings. The engineer shall approve all proposed methods of jacking, loading, winching, etc. prior to the work being undertaken.

Unloading and erection of the structural steel girders shall be under the direction of a Professional engineer, registered or licensed to practice in the Province of Manitoba.

The Professional engineer shall be experienced in bridge girder erection and be present for all stages of the girder erection. Loose timber blocking will not be permitted for use as temporary works for any aspect of girder erection. It is the Contractor's responsibility to ascertain the actual weight of the girders.

2. Equipment

All cranes, rigging and equipment shall be in good condition and properly maintained at all times during the period of the work. The engineer shall, at his/her discretion, verify capacity and state of equipment provided and any equipment found not meeting the requirements for erection work shall be removed and replaced. Slings and other lifting devices that will be in contact with the structural steel work shall be of a type which shall not damage painted surfaces.

3. Erection

The engineer shall be notified in writing of the starting date at least two weeks prior to the commencement of field operations. Work shall not be carried out until the engineer is on the site.

Components shall be lifted, placed and maintained in position using appropriate lifting equipment, temporary bracing, guys or stiffening devices so that the components are at no time overloaded, unstable or unsafe.

Additional permanent material may be provided, if approved by the engineer, to ensure that the member capacities are not exceeded during erection. Release of temporary supports or temporary members, etc. must be gradual and under no circumstances will a sudden release be permissible.

Unless otherwise approved by the engineer, at least 50% of the holes in the joints shall be filled with drift pins or hand tightened bolts prior to removing the crane.

At least 50% the bolts required in the flanges shall be installed. For roadway or railway overpass structures, drift pins shall not be left in place over traffic when the crane is removed.

For temporary fit ups, main girder splices and connections shall be aligned with drift pins and a sufficient number of fitting up bolts shall be installed to maintain the integrity of the connection. The fitting up bolts may be the high strength bolts used in the installation. Drift pins shall be 1 mm larger in diameter than the required bolts.

Excessive drifting that distorts the metal and enlarges the holes is not allowed. Reaming up to 2 mm over the nominal hole diameter is permitted, except for oversize or slotted holes. Repairs to erected material will only be permitted after the repair procedure has been approved by the engineer.

Filling of misplaced holes by welding is permitted only with the written approval of the engineer. Material intended for use in the finished structure shall not be used for erection or temporary purposes unless such use has been shown on the Shop Drawings, erection diagram or authorized by the engineer.

Hammering that will damage or distort the members is not permitted. Surfaces that will be in permanent contact shall be cleaned immediately prior to assembly.

4. Temporary Stresses

The Contractor shall assume full responsibility for ensuring that all bridge member and component stresses are within permissible limits at all stages of the construction work.

The Contractor shall provide all necessary additional steel reinforcement, bracing or other measures required to ensure that the erection procedures do not over stress any temporary or permanent member or component at any stage of the Work.

5. Alignment and Camber

The structural steel girders shall be erected to the proper alignment in plan and in elevation, taking into account the dead load camber shown on the Drawings.

Members shall be aligned to the dimensional tolerances specified in CAN/CSA W59-M, but in no case, shall it deviate by more than 50 mm from the theoretical location.

Alignment shall be measured from survey lines joining the ends of any test length of a member.

6. Temporary Bracing

The contractor shall be responsible for the design, supply, installation and removal of erection bracing, temporary wind bracing, lateral stability bracing and longitudinal ties as may be required during and immediately following the erection of structural steel girders.

The bracing shall be designed and installed so that it will not interfere with the installation of steel diaphragms.

7. Lifting Devices

After the engineer has approved the erection positions of the girders, all lifting devices shall be removed to the satisfaction of the engineer.

5.7 Equipment for Dredging, Trenching and Tunneling

Dredging Necessity

In order to maintain the required level of water in a harbour the external materials deposited in the bed has to be removed from time to time. This operation of removal of materials from the sea or river bed is called dredging. The mechanically operated equipment used for this purpose is called as dredger. Dredging has to be done periodically for the reasons noted below:

- During the construction of a structure in a harbour complex, the wastes and other construction rubbish gets deposited on the bed. This results in the reduction in the available depth of water for berthing of vehicles. Thus, to maintain the design depth the excess material has to be removed.

- Initially, before finalizing a harbour site, the depth of water required has to be decided. If the site does not provide this, a dredge has to be done.

- Waves and tides have a tendency of depositing sand and silt.

Types of Dredgers

Some of the popular dredgers are discussed below:

1. Bucket or Ladder Dredger

This consists of a chain of buckets fixed to continuous elevator belting.

It comprises of:

- Cable arrangement for lowering or raising the ladder.

- Wheel for manually or mechanically operating the chain of buckets.

- Containers for receiving the dredged material.

2. Hydraulic or Cutter Dredger

A hydraulic dredger has a twofold arrangement, viz., operation of dredging and transporting the dredged material to the disposal site through discharging pipe.

In case of loose materials, there is no need for dredge but the material is removed by suction. Hard materials such as stiff clay or dense sand, a rotating cutter is used at the end of the suction pipe.

Water jet is used for loosening dense or stiff materials instead of cutting. This type of dredger is also called as sand pump. This type of dredger is even capable to cut bed rocks.

3. Grab Dredger

This type of dredger consists of a "grab" which is suspended by a cable. The grab is extended by operating the boom of a crane.

Steam, diesel or electric power is used for operating grab cranes. The dredgers may have one to four grab cranes mounted on the dredger.

Special cutting teeth are provided with the grabs for cutting hard rocks. In order to remove cut rock materials the grab attachment is replaced by the poly–grab.

4. Dipper Dredger

This dredger consists of a pontoon fitted with a dipper shovel. The bucket has a shovel which has a striking end. This is used for striking the strata to be dug.

When the hoist is released the dipper sticle goes down and the teeth of the bucket bite into the soil.

When the hoist is pulled up the dipper sticle along with the bucket containing the dug material is pulled up. It is then swung towards the position decided to be dumped.

5. Rock Dredger

This consists of a pontoon to which a dropping hammer of weight 6 to 10 tons is attached. The hammer is allowed to fall, which breaks the rock and the frequency of impact breaks more of the rock. The broken material is removed. The operations are done by steam or diesel engine fitted to the pontoon.

The dipper dredger is classified based on the bulk capacity. It is generally of 1 to 3 cubic metre and large capacity dredgers up to 12 cubic metre capacities are available.

6. Hopper Barge

The dredged material is collected in a hopper fitted to the dredger or discharged into another vessel stationed next to the dredging vessel. This collecting vessel has hoppers for receiving the dredged materials.

The dredged materials are discharged into deep waters or to a shore jetty. In the case of dumping into deep sea, the hoppers are opened and the collected materials fall under gravity.

Trenching

Trenching is done by the equipment called as trenchers. Trenchers are used for excavating trenches or ditches of variable width and depth. The present day trenchers can be used to make trenches of width up to 12 metres and depth up to 3 metres can be excavated in one pass.

Basically a trencher consists of:

- Self-propelled tractor or carrier mounting fitted with crawler or wheel-type running gear.

- An excavating device consisting of several buckets on the periphery of the wheel alternatively several cutters are provided for digging and discharging continuously.

- A provision of conveying the excavated material.

The common type of trencher excavator is the wheel trenchers. It consists of a pair of circular rims whose outside diameters are connected by V-shaped buckets or cutters. The wheel is turned by a chain drive which connects to the power source. The wheel moves to the top position when discharging the material and at the bottom position while digging.

The buckets or cutters on the wheel perform the work of excavating while travelling upward. When the wheel reaches the top position the excavated material drops. The dropped material is carried on conveyors and discharged alongside of the trench to form windrow.

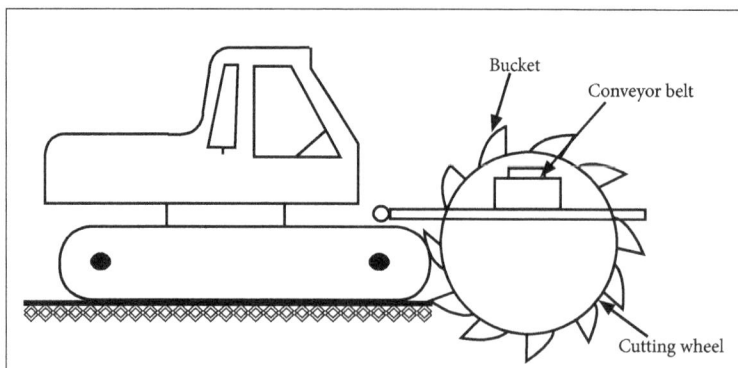

Trencher with tractor.

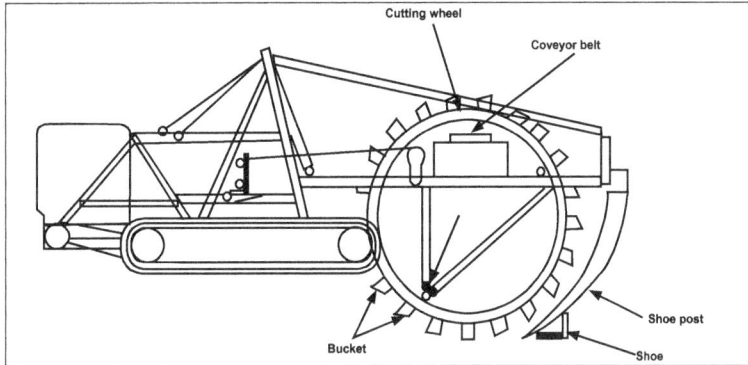

Wheel type ditcher or Trencher.

The windrow is made such that there is some clearance between the windrow and the edge of the trench. The trenches are usually for conveyance of water, gas oil pipeline, telephone cables, drainage, sewers, etc. The selection of trenching equipment depends on various factors, viz., depth and width of the trench, type of soil, disposal of excavated earth, ground water position and the nature of the job.

Permissions

Index

A

Abrasives, 62

Acid-resistant, 4

Addition, 1-3, 6-8, 29, 38, 174, 180

Aggregation, 8

Anode, 149

Atmosphere, 15, 29, 104, 108, 110

Atmospheric Pressure, 16, 148

Automobile, 179

B

Barium, 71

Blocks, 49-51, 54-56, 59, 64-66, 125, 173-174

Boiler, 183, 186

C

Capital Cost, 176

Carbohydrates, 6

Carboxylic Acids, 6

Cast Iron, 23-24

Catalyst, 134

Cathode, 148-149

Cellulose, 103

Chemical Composition, 1, 3

Chemical Reaction, 71

Climatic Conditions, 71

Coefficient, 109, 132

Combustion, 111-112, 186

Composite Materials, 103

Composition, 1, 3, 134

Compressor, 183, 186

Configuration, 15, 63, 79, 174, 190

Conjunction, 119, 137

Constant, 11, 82, 156, 176

Consumer Goods, 165

Cooling Water, 16

Corrosion Resistance, 3, 75

Crude Oil, 108

Cutting Edge, 128-129, 181-182

D

Dams, 3-4, 12, 28, 50, 119, 145-146

Defects, 68, 109

Depth of Cut, 43, 181

Diameter, 18, 20-21, 23, 25, 27, 56, 94-95, 118, 126, 133, 144, 149-150, 197

Displacement, 13, 125-126, 131, 134, 187, 190

Dynamometer, 35

E

E-commerce, 165

Earthquake, 98

Electric Current, 148

Engineering, 1-2, 14, 28, 41, 100, 131-132, 166, 170

Environment, 167

Evaporation, 8-9, 15, 23, 26

Exhaust, 83

F

Fillet, 68

Flexural Strength, 23-24, 36

Floods, 131

Forecasting, 165

Friction, 19, 81, 124, 132, 140, 161, 171, 186, 195

Fuel Consumption, 176

H

Hardness, 32, 92

Humidity, 8, 16, 26, 29

I

Information Systems, 166

Ingredient, 1, 191

Input, 166

Insulation, 3, 50, 58, 65-66, 103, 105, 109, 173

Internal Combustion Engines, 186

M

Maximum Pressure, 36

Means, 13, 17, 34, 43, 63-64, 80, 82, 88, 97, 118, 125, 141-143, 148, 155, 168, 172-173, 178, 191

Methodology, 162

Mild Steel, 56, 95

Motion, 189, 192

O

Odd Number, 190

Organic Compounds, 7

P

Pallets, 166

Penetration, 22, 36, 71, 103, 134, 140

Petrol Engine, 180

Piston, 184-185

Poisson, 23

Polymer, 9

Probability, 29

Q

Quality Control, 28, 34

R

Radius, 14, 68

Rainfall, 105

Reservoir, 119

Resonant Frequency, 35, 186

Resource Allocation, 165

Rotation, 170

S

Stainless Steel, 144

Standardization, 166

String, 94, 101-102

Surface Finish, 10

Surface Tension, 7

T

Tensile Strength, 24-25, 134

Translation, 155, 157

Transportation, 7, 131, 162, 165-166, 171, 174, 190, 192-195

U

Uniform Distribution, 12

V

Vacuum, 108, 148

Variables, 39

Velocity, 8, 29, 33-35, 110, 149

Ventilation, 49, 59, 83, 113

Viscosity, 135

www.ingramcontent.com/pod-product-compliance
Lightning Source LLC
Chambersburg PA
CBHW062000190326
41458CB00009B/2920